Framing Square

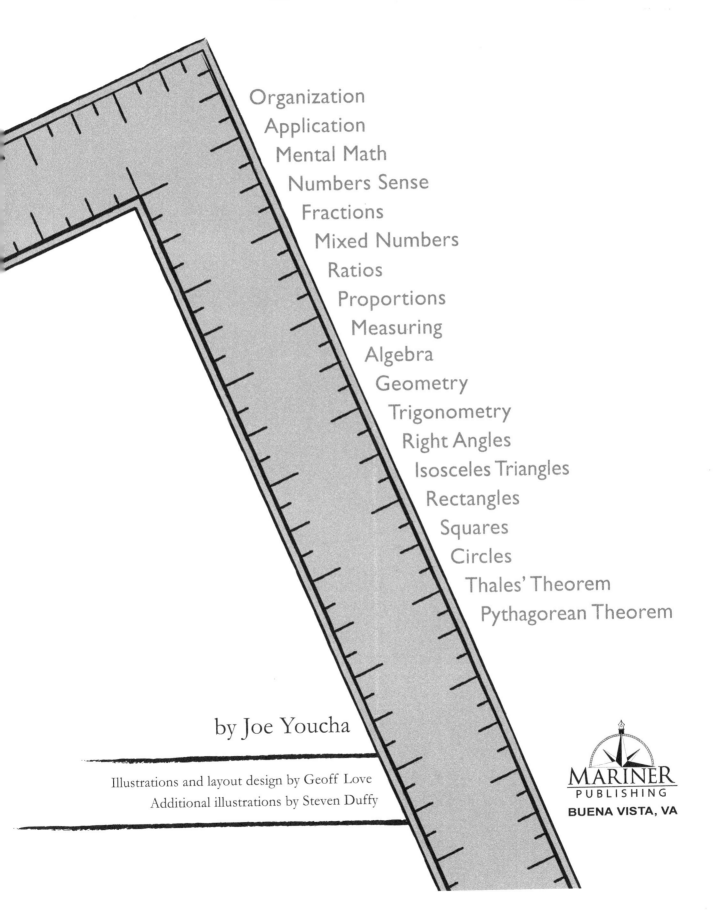

Organization
Application
Mental Math
Numbers Sense
Fractions
Mixed Numbers
Ratios
Proportions
Measuring
Algebra
Geometry
Trigonometry
Right Angles
Isosceles Triangles
Rectangles
Squares
Circles
Thales' Theorem
Pythagorean Theorem

by Joe Youcha

Illustrations and layout design by Geoff Love
Additional illustrations by Steven Duffy

MARINER
PUBLISHING
BUENA VISTA, VA

Copyright © 2017 by Joe Youcha

All rights reserved, including the right of reproduction in whole or in part in any form without the express written permission of the publisher.

1 3 5 7 9 10 8 6 4 2

Library of Congress Control Number: 2017937738
Framing Square Math
By Joe Youcha

p. cm.
1. Education: Teaching Methods & Materials—Mathematics
2. Education: Non-Formal Education
3. Education: Vocational

I. Youcha, Joe, 1962– II. Title.
ISBN 13: 978-0-9975226-4-8 (softcover : alk. paper)

Illustrated by Geoff Love and Steven Duffey
Layout by Geoff Love and Karen Bowen

Mariner Media, Inc.
131 West 21st Street
Buena Vista, VA 24416
Tel: 540-264-0021
www.marinermedia.com

Printed in the United States of America

This book is printed on acid-free paper meeting the requirements of the American Standard for Permanence of Paper for Printed Library Materials.

Dedication

This book is dedicated to John Marconi, a union millwright from Philadelphia (Local 1906), who cared deeply about passing his craft and knowledge to others.

Math Topics

- **Ratios**
- **Proportions**
- **Fractions**
- **Mixed Numbers**
- **Measuring**
- **Geometry**
- **Algebra**
- **Trigonometry**
- **Right Angles**
- **Isosceles Triangles**
- **Rectangles**
- **Squares**
- **Circles**
- **Thales' Theorem**
- **Pythagorean Theorem**

Contents

Introduction   1

    The History of the Framing Square   4

    Right Angles   6

    Powerful Triangles   7

    Parts of a Framing Square   8

    Drawing a Circle (Intro Exercise #1)   9

    Finding the Center of a Circle (Intro Exercise #2)   10

    Why Does This Work?   11

Fractions   13

    Drawing Fractions of a Square   14

    Practicing with Fractions of a Square   15

Measuring   17

    Measuring to 1/16"   18

    Adding Measurements   19

    Subtracting Measurements   20

    Multiplying Measurements   21

    Dividing Measurements   22

    Measuring Using the 1/10th Scale   23

    Adding Decimals Using the 1/10th Scale   24

    Subtracting Decimals Using the 1/10th Scale   25

    Multiplying Decimals Using the 1/10th Scale   26

    Dividing Decimals Using the 1/10th Scale   27

Contents

Measuring Continued

Converting Decimal Inches to Fractions of an Inch	28
Scale Drawing Using the ¹⁄₁₂" Scale	27

Calculating with Ratios and Proportions — **31**

Slipping the Square	32
Multiplying with a Framing Square	33
Dividing with a Framing Square	34
Why Does This Work?	35
Finding Slope	36
Using Slope to Find "Run" Lengths	37
Using Slope to Find "Rise" Lengths	38

Circles and Pi — **39**

Finding Circumference	40
Ways to Figure the Area of Circles	41
Calculating the Areas of Circles	42
Finding Squares and Circles with the Same Area	43
Dimensioning Proportional Rectangles	44
Making an Octagon with ⁷⁄₂₄ths	45

Pythagorean Theorem — **47**

Proving the Pythagorean Theorem	48
Finding the Diameter of a Circle Whose Area is Equal to the Area of Two Given Circles	49

Layout Exercises — 51

- Dividing a Line into Equal Parts — 52
- Dividing a Board into Equal Parts — 53
- Laying Out Polygons Within a Circle — 54
- Dividing a Longer Line into Equal Parts — 56
- Dividing a Tapered Space into Equal Parts — 57
- Laying Out a 10 Degree Angle — 58
- Drawing an Ellipse—Locating the Foci — 59
- Drawing an Ellipse—Pin and String Method — 60

Coordinate Planes and Algebra — 61

- Descartes and Galileo — 62
- Creating the Coordinate Plane and Coordinates — 63
- Locating Points on the Coordinate Plane — 64
- Checking Layout on the Coordinate Plane — 65
- Finding Slope with the Framing Square — 66
- Finding Slope with Algebra — 67
- Linear Equations — 68

Framing Square Trigonometry — 69

- Drawing the Trigonometric Functions (Sine and Cosine) — 70
- Drawing the Trigonometric Functions (Tangent, Secant, Cotangent, and Cosecant) — 71
- Finding Length Using Tangent — 72
- Finding Length Using Cosine — 73

Contents

Framing Square Trigonometry Cont.

    Measuring 10 Degrees Using the Tangent Function   74

Build Your Own Tools   75

    Making Your Own Framing Square   76

    Making a Large Sliding Bevel   77

    Using a Wooden Trammel to Draw an Ellipse   78

Appendices   79

    Carpenter's Math Checklist   80

    Creating a Perpendicular with Thales' Theorem   81

    Finding the Height of an Object Using Similar Triangles   82

    Finding the Height of an Object Using Trigonometry   84

Hands On Math Glossary   87

Introduction

Purpose of this Book
This book targets math teachers who want to teach "hands on":

- Teachers who teach themselves.
- Teachers who teach others.
- Teachers whose students ask, "When am I ever going to use this stuff…"

Hopefully, this book can be used to help their students learn the skills and see that math is useful.

How to Use This Book
A "cookbook" for how to use the Framing Square while teaching hands on math, this book includes exercises, some history, "worked numbers" that make the transition between hands on math and textbook math, as well as a hands-on math glossary. Designed to complement, not replace, a math text, these materials focus on teaching five foundational math skills:

1. *Organization:* These are mostly layout projects that have to be done "step by step." There are no shortcuts. The exercises physically demonstrate the need to proceed from step "A" to step "B" to step "C." The same process applies to working numerical problems.
2. *Recognizing the Application of Math:* Students see situations where math is applied. They then use a tool, with their hands, to do the necessary calculations and constructions.
3. *Mental Math Skills:* Basic math facts are foundational skills. Students are hamstrung if they can't perform basic operations such as multiplication, division, and factoring.
4. *Numbers Sense:* Realizing that numbers can be used to describe the real world motivates students to find the relationships between sets of numbers, as well as between shapes and the numbers that describe them.
5. *Fractions:* Measurements are fractions. Ratios are fractions. Proportions are fractions. The exercises in this book depend upon measurements, ratios and proportions. Fractions don't have to be scary; they're useful and higher math (Algebra, Geometry, and Trigonometry) uses them intensively.

Learning these five foundational skills leads students to increased opportunities in school and work. More importantly, surveys have shown that participating in project based math education (building things while learning math) changes students' attitudes towards math, work, and their own capabilities. These exercises help overcome the fear of basic math skills, as well as the almost irrational fear of Geometry, Algebra, and Trigonometry.

Why a Framing Square?
The Framing Square integrates a range of foundational math skills. With one tool, students can use their hands, eyes, and mind to progress through arithmetic, fractions, decimals, Geometry, Algebra, and Trigonometry. The same mind and hands using the same familiar tool makes progress logical and skill attainment comfortable.

Why These Types of Materials are Needed
Math wasn't widely taught in classrooms until the first half of the 1800s. Yet, math was used to build roads, buildings, ships, and structures at least since the Egyptian pyramids in 2700 B.C.

Introduction

Do the math. A broad calculation says that only about 4 percent of humankind's math education has been in a classroom. The rest has been "on the job," "Hands-On." This method of calculation and instruction successfully taught human beings for thousands of years. This book tries to learn from history and give teachers the opportunity to reach their students through hands-on exercises using one of the most powerful building tools ever developed, the Framing Square.

Math calculations don't have to be rows of numbers lined up on top of one another. They weren't always. Most calculations were done with Geometry before 1840. These exercises present students another way to approach math, which they usually just see as numbers to be plugged into an electronic calculator. Both numeric and geometric calculations demand organization and accuracy. Once learned these skills can be transferred to other applications.

This is visual math—math described through ratios and proportions and usually expressed through similar triangles. Sixty-five percent of students in America are visual learners. Why don't we teach for them?

Which Square?
You can buy a square, or build your own (see page 76). The framing square is an easy to build calculator using the most basic math. All you need are evenly spaced marks laid out on both legs of a right triangle and you can figure out almost anything practical that you need to know.

If you buy a square, buy a good one such as Starrett RS-24, Stanley 45-011, or Empire 1140. This will allow you to do all the exercises. If you build your own square, you won't have the $1/10$ths or $1/12$ths SCALES, and your scales will all be on the outside of the square. This will slightly limit the number of exercises you can do; but, your students will have the experience of building and using their own tool—a wonderful thing. (It's also much cheaper.)

Start Building Your Student's "Eye"!
Craftspeople believe in a "builder's eye." It can be artistic; but that's only part of it. Builders also see the math inherent in the building process and how to use it. They may just not call a skill by the same name a mathematician uses; but a builder's eye is an eye that sees mental math, numbers sense, as well as proportion and organizes them into an application that produces something—hopefully something useful!

Resources
As a teacher, you'll need to work through these exercises. They are written for you. They also are laid out to be used by your students. Broadly, the exercises build upon one another. Not every student will need to do all the exercises. They cover a range of skills and grade levels, and some involve large layout spaces.

Alongside the exercises, we work the numbers to make the connection between math skills as the student uses them in the exercise and math skills as they see them in a textbook. We also encourage you to use the hands-on math Glossary at the end of this book. Each term in the Glossary is in **BOLD SMALL CAPS** in the exercises. The illustrations are representative, don't try and read the SCALES on the Framing Squares. The measurements are called out instead. *At the end of most exercises, opportunities for further skill practice are suggested in italics.*

The exercises in the book are meant to introduce a concept, or jog the students' memory. We're assuming they've seen these skills before; or, they

Introduction

are seeing them concurrently with other materials. If the student is learning these math skills for the first time, there should be an accompanying math text, like the Carpenters' Union *Math For The Trades* book. Or, go online to a resource such as Khan Academy. www.kahnacademy.org

This book is about getting back to math basics. It's part of Building To Teach: www.buildingtoteach.org. It ties closely with the United Brotherhood of Carpenters International math training efforts. And, it's closely related to Jim Tolpin and George Walker's work to reintroduce Artisanal Geometry: www.byhandandeye.com.

For more information about this book and Building To Teach contact:
Joe Youcha
jyoucha@buildingtoteach.org
www.buildingtoteach.com

Needed Tools and Materials (Read through the book to see where the following are used.)

- Framing Square with $\frac{1}{16}$th, $\frac{1}{12}$th and $\frac{1}{10}$th Scales (models Starrett RS-24, Stanley 45-011 or Empire e1190)
- Drawing Surface
- Masking Tape
- Pencil
- Paper
- Tape Measure
- Stranded Wire (as for hanging pictures)
- Notebook: to build their own Journal (ideally, a composition notebook with graph paper)
- Finish Nails
- Hammer
- Dividers/Compass
- Chalk Line
- Random-sized wooden blocks
- Plumb Bob (For Outdoor Exercises)
- Level (For Outdoor Exercises)

The History of the Framing Square

Squares makes RIGHT ANGLES—the fundamental component of most building shapes. Over thousands of years, the square evolved into the Framing Square, or "Steel Square" with rulers, or SCALES, along its edges. Until recently, it was the most powerful tool a carpenter used to layout buildings.

Squares started out being made from wood. They were used by the Egyptians almost 5,000 years ago to build the pyramids. Roman aqueducts, roads, buildings, and furniture were built with wooden and metal squares. Wooden squares were used on European timber-framed houses and cathedrals in the Middle Ages.

Building trades' guilds evolved partially to pass down how to use the square. That valuable knowledge largely involved the math skills necessary for construction.

"To a skilled journeyman in our trade, the Framing Square is almost indispensable and a very versatile tool." So starts an old Carpenters' Union training manual on the use of the steel square.

These tools helped generations of skilled craftsmen perform the layout and calculations that built almost anything. Today, the square retains its power; but now, we are applying that power to math education.

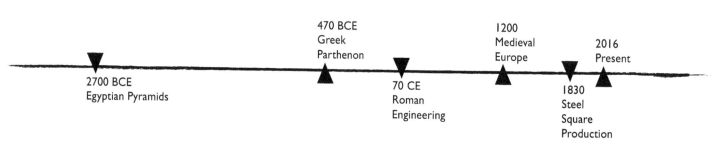

Originally, the SCALES on a square were just evenly spaced marks along the tool's edges. In Europe, the definition of a "foot" depended upon the size of a King's foot. An inch depended upon the width of that King's thumb.

It wasn't until the 1800s that MEASUREMENTS became standardized. Around this time mass production of many different versions of steel squares began. Each had different capabilities and functions. They were such powerful tools that each version seems to have had books written about it.

One of those books calls the Framing Square "the most versatile implement in existence [that] can be used to solve most of our daily problems plus those *rare* layout procedures."

The "problems" they were talking about were math problems. The layout that can be done with the tool truly is refined and subtle, or "rare," as that book says.

Building Math.
Practical Math.
Hands-On Math.

Whatever you want to call it, this is useful math that human beings have been learning and utilizing for hundreds of years.

Framing Square Math

Right Angles

Using the square teaches **Ratios**, **Proportions**, **Decimals**, **Fractions**, **Mixed Numbers**, **Measuring** and practical **Geometry**, as well as a fair amount of **Algebra** and **Trigonometry**.

It all starts with **Right Angles**. Right Angles go by many names… "90 degrees," "square," "normal," "**perpendicular**," as well as "right." It's the "right" angle because at any other **angle** to **level** wood isn't totally in compression. Buildings fell down sooner if their supports were built at those other "wrong" angles.

In Ancient Greece, the Pythagoreans built a religious cult largely around right angles and all that can be done with them. It involved everything from math, to music, to science, and even philosophy. The concepts were so magical, they had to be gifts from the gods.

A core of Geometry, the right angle also creates the **coordinate plane** and describes **slope**—both ideas are foundational to Algebra.

Basic Trigonometry depends on the ratios formed by the lengths of right triangles' sides.

Right angles connect the fundamentals of most math subjects. One of the most useful tools in practical math, they provide the driving force behind the square.

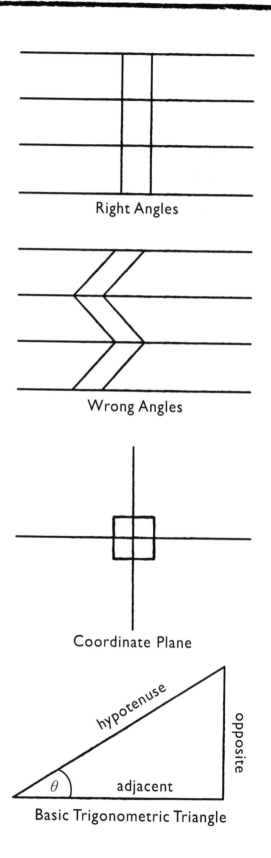

Right Angles

Wrong Angles

Coordinate Plane

Basic Trigonometric Triangle

Framing Square Math

Powerful Triangles

Similar Right Triangles

A lot goes on inside right triangles. For instance, the lengths of their sides form RATIOS. Linking these elegant triangles together into SIMILAR TRIANGLES increases the power of the square. The different triangles have the same interior ANGLES, which makes the lengths of their sides PROPORTIONAL.

Usually, if you know two out of three things, you can determine the third. If you know three out of four things, you can determine the fourth. This can be done with ALGEBRA. Or, we can use the Framing Square and similar triangles to set up the desired PROPORTIONS and "solve for the unknown."

Isosceles Triangles

Closely related to RIGHT ANGLES, ISOSCELES TRIANGLES are one of the most useful tools on the job. One isosceles triangle makes two SYMMETRICAL right angles. When building, they are used to create PERPENDICULARs and symmetrical layout.

Math Predicts the Future

- Will the piece of wood we're cutting fit into what we're building?
- Will those furniture pieces form perfect SQUARES, or RECTANGLES, when assembled?
- Will a roof hold up under a load of snow?

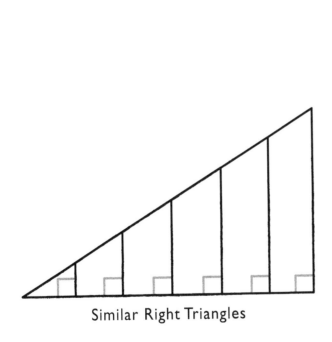

Similar Right Triangles

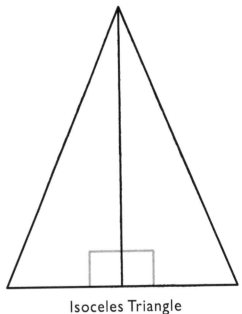

Isoceles Triangle

Framing Square Math

Parts of a Framing Square

Language evolved around trades and their tools. The Framing Square has its own vocabulary. There has never, ever, been a single, standard layout of the square; but there are common names for its parts.

- *Tongue* – The short leg of the square
- *Body* – The long leg of the square
- *Face* – The side of the square showing when the tongue is to the right and the body is pointing down
- *Back* – The side of the square showing when the tongue is to the left and the body is pointing down
- *Scales* – The rulers along the front and back edges of the square. The following is typical:
    - $1/8$" – Face
    - $1/16$" – Face, Back
    - $1/12$" – Back, outside tongue, body
    - $1/10$" – Back, inside tongue

Try to get a square with these scales.

Complementary Tools

There are tools that complement the square and make it even more powerful and easier to use.

- *Sliding Bevel* – A tool that holds the square and has an adjustable blade that creates different angles. You can build this tool with the exercise on page 77.
- *Stair Gauges* – Buttons that can be fastened onto the tongue and body of a square to repeat triangles. They're great for laying out stairs.

You can also build a square out of cardstock. You won't be able to do all the exercises in this book; but you'll be able to do a lot (see page 76).

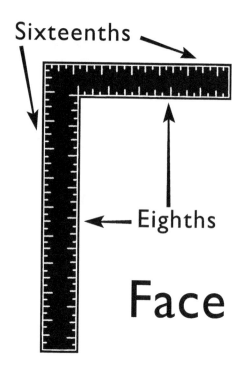

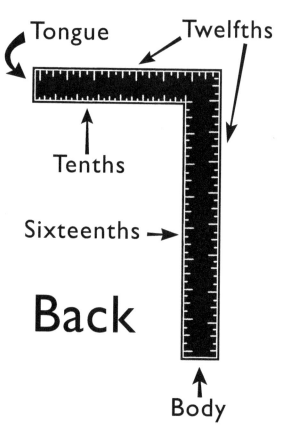

Framing Square Math

Drawing a Circle (Intro Exercise #1)

Let's start off with a couple of introductory exercises to get used to manipulating the square. And, let's do something unusual.

Typically, we use a compass to draw a CIRCLE. Most compasses can't draw a circle bigger than 8" in DIAMETER. Bigger circles need to get drawn on the job all the time to cut holes or layout columns. A typical Framing Square can draw up to 18" diameter circles. Here's how.

You'll Need:
- A surface into which you can nail: plywood, drywall, etc.
- Finish nails (any size from 4d to 10d will work, you only need two to draw a circle).
- Framing Square
- Hammer
- Pencil

To Start:
- Determine the diameter of your circle.
- Draw a line on your board that will serve as a diameter of your circle (Fig. 1). Make sure you leave enough room on your board to draw the circle (Fig. 2). And remember, the size of the circle is limited by the size of the square.
- Mark the diameter on the line, points A and B in Figure 1.
- Put nails at those marks. The nails just have to be driven in far enough to be stable.
- Lay the square so that the inside of the TONGUE and BODY are touching the nails.
- Place the tip of the pencil in the inside corner of the square, Point "C" in Fig. 2.
- Keeping contact between the inside edges of the square and both nails, swing the square to create half the circle.
- Flip the square onto the other side of the diameter to draw the second half of the circle.

Fig. 1

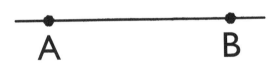

Fig. 2

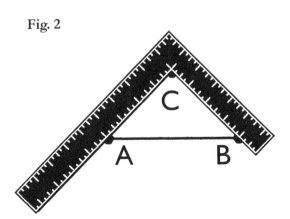

Fig. 3

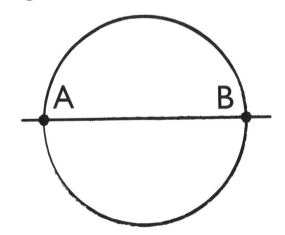

Framing Square Math

Finding the Center of a Circle (Intro Exercise #2)

Now, let's find the CENTER of a CIRCLE using a Framing Square. If we think about the previous exercise, we'll realize that the legs of a RIGHT ANGLE will cross a circle at a DIAMETER—if the VERTEX of that right angle also touches the CIRCUMFERENCE of the circle. And we know two diameters only cross at the center of a circle.

You'll Need:
- Paper, plywood, poster board, or any flat material on which a circle can be drawn.
- Framing Square
- Compass
- Any round, traceable object which can be used to draw a circle (a paint, big soup, or coffee can works well). This ensures there is no visible center point.
- Pencil
- Eraser (mistakes can happen)

To Start:
- Draw a circle, preferably between 6" and 12" in diameter, using your traceable circular object.
- Place the corner of the framing square on a point at the top of the circumference of the circle (Fig. 1).
- Mark where the legs of the square cross the circumference of the circle at points A and B.
- Connect points A and B with a line. This is your first diameter.

Make your second diameter.
- Rotate the vertex of the square around the circle approximately 90 degrees (Fig. 2).
- Mark the points C and D where the legs now cross the circle's circumference.
- Connect points C and D with a line—your second circumference.

Your two diameters cross at the center of the circle point E.

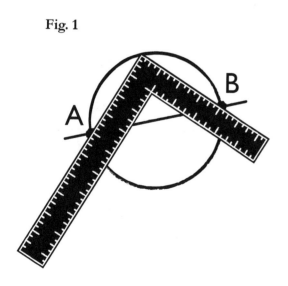

Fig. 1

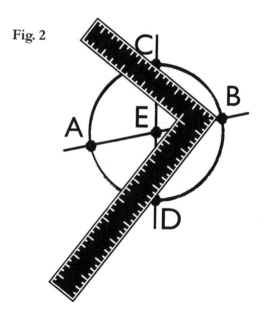

Fig. 2

Framing Square Math

Why Does This Work?

Ancient Greek mathematicians abound. We usually hear about Pythagoras (see later exercises); but who was Thales?

Some people, including Aristotle, said that Thales of Miletus, who lived from 624 BC to 546 BC, was the first Greek philosopher, scientist, and mathematician—although his occupation was that of an engineer.

Going back to our previous two exercises, **THALES' THEOREM** says that if **A**, **B**, and **C** are points on the CIRCUMFERENCE of a circle where the line **AC** is a DIAMETER of the circle, then the ANGLE ∠ **ABC** is a right angle.

Thales' Theorem uses three geometric components: a circle's diameter, its circumference, and a right angle. If we have two components, we can generate the third—just like any three-part equation.

The Framing Square has been our right angle. In Introductory Exercise 1, we used it and the circle's diameter to create the circle. In the second introductory exercise, we used the right angle and the circle's circumference to draw diameters.

It's not how you usually draw a circle; but we're going to see many alternative ways of doing things as we work our way through these exercises.

Can you figure out how to draw a right angle with a compass and straightedge using Thales' Theorem?
(See page 81 in the Appendices for the answer.)

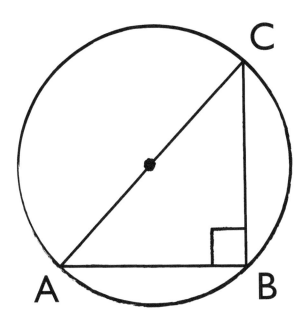

Framing Square Math

11

Fractions

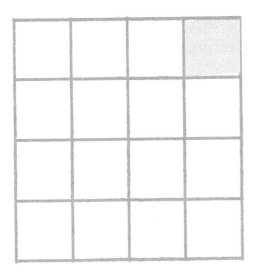
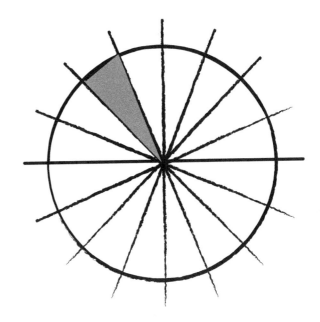

"One Sixteenth"

$\frac{1}{16}$

FRACTIONS are so useful. At the heart of a lot of math, fractions empower ratios and proportions, ALGEBRA, GEOMETRY, and TRIGONOMETRY. Those math skills enable us to build, travel, and live healthy lives. Yet, fractions are just division problems. The "fraction bar" is just a division sign.

$$\tfrac{1}{16} = 1 \div 16 = 16\overline{)1}$$

Misunderstood, underestimated, bypassed by many people, fractions get avoided. Much of this behavior has to do with language. The words describing fractions aren't in English. They're Latin, the language of the Romans.
- *Integer* – Our word for number means "whole" in Latin.
- *Fraction* – In Latin means a piece broken off of the whole.
- *Denominator* – Means "The Namer." It's the bottom number of a fraction and tells us how many "parts" are in the "whole." It's the "name" of the unit with which we're working.
- *Numerator* – Is just the "Numberer" or "Counter" in Latin. It's the top number in a fraction and indicates how many pieces of the "unit" the fraction contains.

Fractions and MIXED NUMBERS are keys to measuring. Before we can use the square fully, we have to know how to measure. Before we can measure, we have to be able to use fractions. This is where the basic skill building starts.

Drawing Fractions of a Square

Ranging from the pitch of a roof to measurements, fractions describe many different situations in the building process. This also means that these situations can all help explain fractions. In this series of exercises, we'll use a basic building shape, the simplest rectangle and the namesake of our favorite tool—the SQUARE—to start exploring fractions.

In this case, you'll draw a 4" by 4" square.

You'll Need:
- Framing Square
- Pencil
- Paper

To Start:
- On the inside of the square, draw a 4" by 4" right angle.
- Mark the whole inches on both legs (1 through 4) (Fig. 1).
- Flip the square; line up the 4" marks on the square with the ends of the lines you just drew.
- Draw the resulting 4" right angle to complete the square.
- Mark the whole inches on both legs (1 through 4) (Fig. 2).
- Draw lines between your marks, as shown in Figure 3.

How many parts are contained in the whole square you've just drawn? (Fig. 3)

Talking Fractions: What is the name of this number? Does it go above or below the fraction bar?

Fig. 1
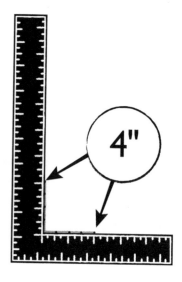

Fig. 2
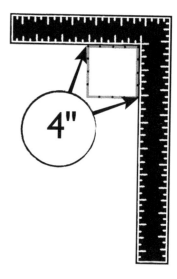

Fig. 3
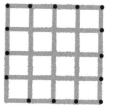

Practicing with Fractions of a Square

Now that you've created the whole square and determined the unit with which you'll work, we can start dividing the square into fractions.

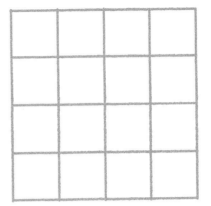

You'll Need:
- The drawing of the square you did in the previous exercise
- Pencil
- Straightedge
- Scissors

To Start:
Only use the lines you've drawn in the previous exercise.

Making Halves
- Using the straightedge and a pencil, determine how many ways you can draw lines that will divide the square into two equal parts.

Making Quarters
- Using the straightedge and a pencil, determine how many ways you can draw lines that will divide the square into four equal parts.

Making Eighths
- Using the straightedge and a pencil determine how many ways you can draw lines that will divide the square into eight equal parts.

As a Challenge...
- *Cut the square out with a pair of scissors.*
- *Figure out if you can fold the square into half, quarters, and eights **without** folding along the lines you've drawn. (Hint: Look at the illustration on page 13)*

Framing Square Math

Measuring

"Measure twice. Cut once."

You only need to do two things to build anything: draw the right line, and cut to it. Drawing the right line depends on **MEASUREMENT**. When you build something, measurement comes first. You can't use the Framing Square to make calculations, or do layouts, without being able to measure.

Measurement tells you the size, or magnitude, of the object. Sometimes, you make measurements directly, with a tick strip, or dividers. Most times, measurements are made with numerical **DIMENSIONS**. We use rulers to create dimensions that approximate the true measurements of the object. Even though the terms get used almost interchangeably, when you're building things, there's a difference between a measurement and a dimension. Measurement is absolute. An object only has one measurement.

The finer the divisions on the ruler, the more accurate the dimension of the object. Using sixteenths is more accurate than using eighths; and using hundredths is way more accurate than using tenths. The square divides inches into $½$, $¼$, $⅛$, $1/16$, $1/12$, $1/10$, and sometimes $1/100$s. You get to pick the unit that allows you to best do your job. This defines "working to the correct **TOLERANCE**."

Why do we use blocks?

In the following exercises, we measure blocks of wood because they are easy to handle, inexpensive, and easy to create. They also mimic so many of the rectangular shapes we build. As **MIXED NUMBERS**, measurements combine whole numbers with fractions. Alongside the exercises, we "work the numbers," so you can see how the "hands-on" math translates to "pencil and paper" math. We also encourage you to use the hands on math Glossary at the end of this book.

Let's start measuring...

Measuring to 1/16"

Let's use a Framing Square to measure the DIMENSIONS of two blocks to the nearest 1/16". Sixteenths are a carpenter's most common unit of MEASUREMENT. A higher TOLERANCE isn't usually needed. We'll record the measurements on the blocks and in a separate table in your notebook.

Fig. 1

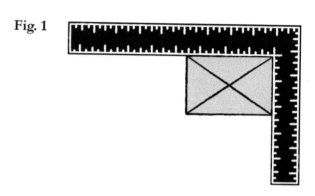

You'll Need:
- Two random-sized blocks, no dimension should exceed 12"
- Masking tape
- Notebook
- Pencil
- Framing Square

Fig. 2

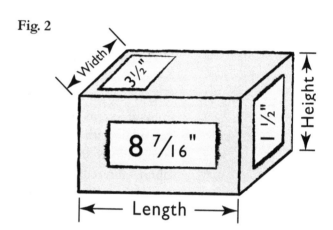

To Start:
- Place 1" long pieces of tape on the side, top and one end face of each block.
- Using the 1/16" scale on the FACE of the Framing Square, measure the longest edge of the first block (the length), and write the number on the appropriate piece of tape (Figs. 1 & 2).
- Measure the next longest edge of the block (the width), and write the number on the appropriate piece of tape (Figs. 1 & 2).
- Measure the next shortest edge of the block (the height), and write the number on the appropriate piece of tape (Figs. 1 & 2).
- Repeat the procedures for the second block.
- Draw a table in your notebook and record the measurements (Fig. 3).

Have a partner check your measurements, while you check theirs.

Fig. 3

	Block 1	Block 2
length		
width		
height		

Adding Measurements

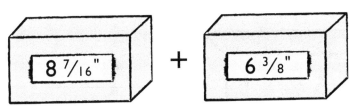
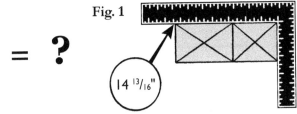

Fig. 1

Combining accurate measurements allows you to build. Let's measure the lengths of two block and add those measurements. At this point, we're "doing math." Don't be scared. Use the Carpenters' Math Checklist (page 80). It prevents a lot of unnecessary mistakes. The "Let's do the Numbers" section to the right leads you through the process. Graph paper also helps keep your work organized.

Adding measurements depends upon doing four things:
- Finding **COMMON DENOMINATORS**
- Adding the **WHOLE NUMBERS**
- Adding the **FRACTIONS**
- Combining the Fractions and Whole Numbers into the answer

You'll Need:
- Notebook
- Pencil
- Masking tape
- Framing Square – Use the 1/16th" scale
- Random sizes of blocks so that each participant has at least two blocks of different sizes

To Start:
- Take two blocks and measure their individual lengths.
- Record the measurements on the piece of masking tape that has been put on each block.
- Draw a table in your notebook and enter the lengths. (Fig. 2)

Fig. 2

	Block 1	Block 2
length		

- In your notebook, add the lengths together and calculate the answer. Remember to find Common Denominators.
- Measure the two blocks together to obtain the actual combined length. (Fig. 1)
- Compare the calculated measurement to the actual measurement.

If you're working in a group, for more practice, try removing the tape from one block and exchanging that block with another participant. Now, repeat the steps with the new pair of blocks.

Let's do the Numbers...

$8 \frac{7}{16} + 6 \frac{3}{8} = ?$

To find the Common Denominator:

Ask, "Which is smaller: $\frac{1}{16}$ or $\frac{1}{8}$?"

The answer is: "$\frac{1}{16}$"

So, we work in $\frac{1}{16}$ths, which is our common denominator. To convert $\frac{3}{8}$ths from $\frac{1}{8}$ths to $\frac{1}{16}$ths without changing its value, multiply $\frac{3}{8}$ by $\frac{2}{2}$ (an equivalent of 1).

$\frac{3}{8} \times \frac{2}{2} = \frac{6}{16}$

Add whole numbers: $8 + 6 = 14$

Add Fractions: $\frac{6}{16} + \frac{7}{16} = \frac{13}{16}$

Combine whole numbers and fractions: $14 \frac{13}{16}"$

Framing Square Math

Subtracting Measurements

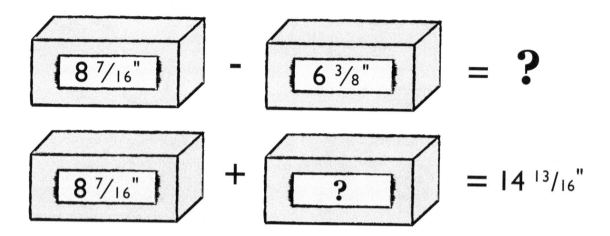

After addition comes subtraction. Builders constantly take lengths of materials, cut them into smaller pieces and then combine them into a structure. In this exercise, you'll accurately subtract the length of the smaller block from the measurement of the larger block; and, subtract the length of one block from the combined measurement of the two blocks. Remember to use the Carpenter's Math Checklist! (p. 80)

You'll need:
- The tools and materials you used in the "Adding Measurements" exercise.

To Start:
- In your notebook, create a table with columns for the lengths of both blocks and their combined length (Fig. 1).
- Measure the combined length of two blocks, and record the measurement.
- Measure the length of one block only, and record the measurement.
- Subtract the length recorded in the previous step from the combined lengths of the blocks to obtain the length of the second block.
- Record the resulting calculation.
- Measure the second block and compare the actual measurement to the calculated measurement

Find more blocks and keep measuring until you're good at adding and subtracting measurements.

Fig. 1

	Block 1	Block 2	Combined
length			

Let's do the Numbers...

$8\,7/16 - 6\,3/8 = ?$

$3/8 \times 2/2 = 6/16$	Find the Common Denominator.
$8 - 6 = 2$	Subtract whole numbers.
$7/16 - 6/16 = 1/16$	Subtract fractions.
$2\,1/16"$	Combine whole numbers and fractions for the answer.

Multiplying Measurements

If you were working on a job, you're almost certain to estimate materials, or calculate the weight of something. When doing these tasks, measurements get multiplied all the time.

Multiplying two DIMENSIONS (length x width) yields a surface's area in "SQUARE" units of measure. In our case we'll end up with square inches.

Multiplying three dimensions (length x width x height) yields an object's volume in "CUBIC" units of measure. In our case, we'll end up with cubic inches. Multiply that volume by the correct material coefficient and you'll have that object's weight.

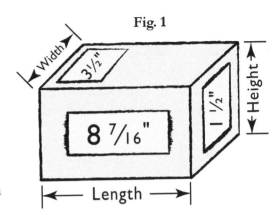

Fig. 1

You'll Need:
- Pencil
- Notebook
- Masking tape
- Random-sized blocks
- Framing Square

Fig. 2

	Block 1	Block 2
length		
width		
height		

To Start:
- Place 1" pieces of tape on the side, top, and one end face of the block (Fig. 1).
- In your notebook, create a table with columns for the lengths, widths, and heights of each block (Fig. 2).
- Measure and record the length and width of the side face of a block, and write the number on the appropriate piece of tape and in the table.
- In your notebook, multiply the length by the width, as in "Let's do the Numbers" to the right.

Remember to convert the MIXED NUMBERS into IMPROPER FRACTIONS and then convert your results back into mixed numbers. Also, use the Carpenter's Math Checklist! (p. 80)
- Measure and record the height and width of the end surface of the same block.
- Calculate the end surface's area by multiplying its height and width and record the result in your journal.

Repeat the process using the second block. Calculate the volume of your blocks. Calculate the weight of your block. See info on page 26.

Let's do the Numbers...

$8\,7/16 \times 3\,1/2 = ?$

$8\,7/16 = 135/16$ — Convert $8\,7/16$ to an improper fraction

$8 \times 16 = 128$ — Multiply the whole number by the Denominator.

$128 + 7 = 135$ — Add the resulting number to the original Numerator.

$135/16$ — Place the result over the original Denominator.

$3\,1/2 = 7/2$ — Convert $3\,1/2$ to an improper fraction

$3 \times 2 = 6$ — Multiply the whole number by the Denominator.

$6 + 1 = 7$ — Add the resulting number to the Numerator.

$7/2$ — Place the result over the original Denominator.

$135/16 \times 7/2 = ?$ — Multiply improper fractions.

$135 \times 7 = 945$ — Numerator times Numerator.

$16 \times 2 = 32$ — Denominator times Denominator.

$\dfrac{945}{32} = 32\overline{)945}$ — Simplify by division.

$$32\overline{)945} \quad \text{29 R 17}$$

The remainder is in 1/32ths.

$$\begin{array}{r} 64 \\ \overline{305} \\ 288 \\ \overline{17} \end{array}$$

The Answer is $29\,17/32$ square inches

Framing Square Math

Dividing Measurements

Measurements also get divided all the time to estimate, or calculate, **PERIMETER**, area, volume, and weight. Determining the even spacing of fastenings on a piece of furniture, or the spacings of structural columns in a building, all depend on dividing measurements.

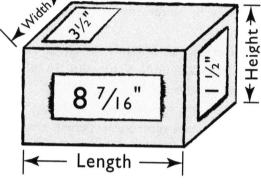

Fig. 1

You'll Need:
- Pencil
- Notebook
- Masking tape
- Random-sized blocks
- Framing Square
- Graph paper

Fig. 2

	Block 1	Block 2
length		
width		
height		

To Start:
- Place 1" pieces of tape on the side, top and one end face of each block.
- Measure and record the length, width, and height of the side face of a block; and write the numbers on the appropriate pieces of tape.
- In your notebook, create a table with columns for the lengths, widths, and heights of each block (Fig. 2).
- In your notebook, divide each dimension by four, as shown to the right in "Let's do the Numbers." Remember to use the Carpenter's Math Checklist (p. 80).
- Remember to convert the **MIXED NUMBERS** into **IMPROPER FRACTIONS** and then convert your results back into mixed numbers. Also, remember the "Keep, Change, Flip" rule for dividing fractions.
- Calculate and record the result in your journal.

One way to think of this question is to say that the length of the block is made up of one hundred and thirty five $1/16$" segments. Then ask, "What is one fourth of those one hundred and thirty five $1/16$" segments?"

Repeat the process using the second block.

Let's do the Numbers...

$8 \, 7/16 \div 4 = ?$

$8 \times 16 = 128$ — Convert mixed numbers to improper fractions.

$128 + 7 = 135$

$8 \, 7/16 = \dfrac{135}{16}$

To divide fractions, remember: "keep, change, flip."

$\dfrac{135}{16}$ — Keep the first number

$\dfrac{135}{16} \times$ — Change the sign to the inverse math operation.

$\dfrac{135}{16} \times \dfrac{1}{4} =$ — Flip (invert) the second number.

$\dfrac{135}{16} \times \dfrac{1}{4} = \dfrac{135}{64}$ — Solve the multiplication problem.

$64 \overline{)135}$ → 2 R 7, $\dfrac{128}{7}$ = $2 \, 7/64$ — Simplify back into a mixed number.

Framing Square Math

Measuring Using the ⅒" Scale

Many trades, such as millwrights, machinists, and mechanics, perform calculations in **DECIMALS**. Almost anytime you use an electronic calculator to manipulate measurements, you'll be working in decimals. The Framing Square also can calculate in decimals. Measuring, as usual, is the first step.

In this exercise, you'll use the ⅒" scale on a Framing Square to measure the wood blocks in **DECIMAL INCHES**.

The ⅒" scale should be located on the inside of the **TONGUE'S BACK**.

You'll Need:
- Notebook
- Block of wood
- Masking Tape
- Framing Square with ⅒th scale
- Pencil

To Start:
- Place 1" pieces of tape on the side, top and one end face of the block. (Fig. 1).
- Create a table in your notebook, like the one shown, into which measurements can be entered (Fig. 3).
- Measure the length, width, and height of the block to the nearest ⅒".
- Write these measurements as decimals on the tape, as well as in your table.

Repeat the process using the second block.

Fig. 1

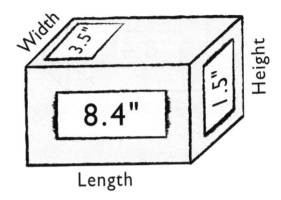

Fig. 2

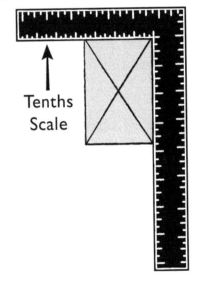

Fig. 3

	Block 1
length	
width	
height	

Framing Square Math

Adding Decimals Using the 1/10" Scale

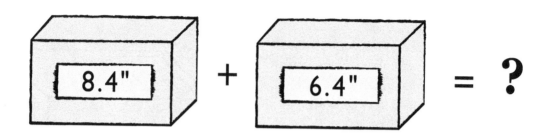

Let's use the 1/10th scale on a Framing Square to add DECIMAL inch measurements.

You'll Need:
- Notebook
- Random-sized blocks of wood
- Masking Tape
- Pencil
- Framing Square with 1/10th scale

To Start:
- Place 1" pieces of tape on the side, top and one end face of the block.
- In your notebook, create a table into which measurements can be entered (Fig. 2).
- Measure the length of each block to the nearest 1/10" (Fig. 1).
- Record the measurements, as decimals, on the piece of masking tape that has been put on each block.
- Also, write these measurements into the table.
- Add the lengths together, as in "Let's do the Numbers."
- Check calculations by physically measuring the blocks.

If you'd like more practice, add the widths and heights of the blocks using the table.

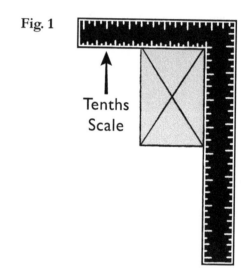

Fig. 1 — Tenths Scale

Fig. 2

	Block 1	Block 2
length		
width		
height		

Let's do the Numbers...

8.4 + 6.4 = ?   Set the problem vertically.

```
  8.4
  6.4
 ----
 14.8
```
Line up the decimal points

Subtracting Decimals Using the ⅒" Scale

In this exercise, you'll use the ⅒th scale on a Framing Square to subtract **DECIMAL** measurements. We'll do this three different ways.

You'll Need:
- Notebook
- Random-sized blocks of wood
- Masking Tape
- Pencil
- Framing Square with ⅒" scale

First, let's subtract the total width from total length.

To Start:
- Place 1" pieces of tape on the side, top and one end face of the block (Fig. 1).
- Draw the table on the left in your notebook (Fig. 2).
- Measure the longest edge of each block (the length) using the ⅒ths scale on the Framing Square.
- Write the measurement in decimal form to the nearest ⅒th of an inch in the appropriate place on the block and in the table.
- Measure the next longest edge of each block (the width) and record.
- Measure the next shortest edge of each block (the height) and record.

In your notebook, perform the following operations for each block:
- Subtract the width from the length.
- Subtract the height from the width.
- Subtract the height from the length.

Have a partner check your work and do the same for them.

For more practice, do the "Subtracting Measurements" exercise on page 18 using the ⅒" scale, rather than the 1/16" scale.

Fig. 1

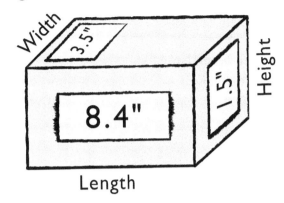

Fig. 2

	Block 1	Block 2
length		
width		
height		

Let's do the Numbers...

8.4 - 3.5 = ?   Set the problem vertically.

```
  8.4
- 3.5
-----
  4.9
```
Line up the decimal points

Framing Square Math

Multiplying Decimals Using the ⅒th Scale

Now, we'll use the ⅒th scale on a Framing Square to measure wood blocks, multiply the measurements and calculate area, volume, and weight in DECIMAL SQUARE and CUBIC inches.

You'll Need:
- Notebook
- Pencil
- Random-sized blocks of wood
- Masking Tape
- Framing Square with ⅒" scale

To Start:
- Place 1" pieces of tape on the side, top and one end face of the block (Fig. 1).
- Draw a table in your notebook similar to the one shown (Fig. 2).
- Measure the longest edge of each block (the length) using the ⅒th scale on the Framing Square.
- Write the measurement in decimal form to the nearest ⅒th of an inch in the appropriate place in the table and on the block.
- Measure the next longest edge of each block (the width) and record the measurement.
- Measure the next shortest edge of each block (the height) and record the measurement.
- Multiply the length by the width for each block to find the area of that particular block face.
- Multiply that area by the block's height to find it's volume.

Calculating Weight
With a coefficient, volume can be used to calculate weight. Our block is made of wood. Wood weighs about .02 pounds per cubic inch.
- Multiply our block's volume by .02.
- How much should our block weigh?

Lead has a weight of about .4 pounds per cubic inch.
- *How much should our block weigh if it were made from lead?*

How much does your block weigh when made from: Wood? Lead?

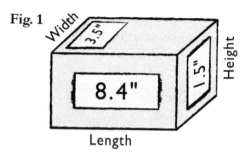

Fig. 1

Fig. 2

	Block 1	Block 2
length		
width		
height		

Let's do the Numbers...

8.4 × 3.5 = ?   Set the problem vertically.

```
  8.4
  3.5
  ---
  420
 2520
 ----
 2940
```
Multiply the length by the width.

```
  8.4
  3.5
```
Count the decimal places in the equation (two places).

```
 29.40
```
square inches

Use the number of decimal places in the equation to locate the decimal point in the answer.

```
 29.40
   1.5
 -----
 14700
 29400
 -----
 44.100
```
Multiply the area by the block's height (follow the previous procedure).

44.10 cubic inches

Now, calculate the approximate weight of the wooden block.

```
 44.1   cubic inches
x .02   lbs/cubic inch
 ----
 .882   lbs
```
Our block should weigh less than a pound.

Dividing Decimals Using the ¹⁄₁₀th Scale

We're measuring in ¹⁄₁₀" scale. Sometimes machinists and millwrights measure in ¹⁄₁₀,₀₀₀" scale. Those are the types of TOLERANCEs demanded by very accurate work. Precision spacing of holes, or the thickness of shims, can affect the operation of machines worth millions of dollars. Accuracy depends upon the division of DECIMALs (which is the same process you're doing here).

You'll Need:
- Notebook
- Pencil
- Random-sized blocks of wood
- Masking Tape
- Framing Square with ¹⁄₁₀" scale

To Start:
- Place 1" pieces of tape on the side, top and one end face of the block (Fig. 1).
- Draw a table in your notebook similar to the one shown (Fig. 2).
- Measure the longest edge of each block (the length) using the ¹⁄₁₀th scale on the Framing Square.
- Write the measurement in decimal form to the nearest ¹⁄₁₀th of an inch in the appropriate place on the block and in the table.
- Measure the next longest edge of each block (the width) and write the measurement in the appropriate places on the block and in the table.
- Measure the next shortest edge of each block (the height) and write the measurement in the appropriate places on the block and in the table.
- Divide the length by the width for each block.
- Divide the resulting products by the height for the each block.

Fig. 1

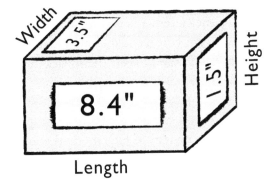

Fig. 2

	Block 1	Block 2
length		
width		
height		

Let's do the Numbers...

8.4 ÷ 3.5 = ?   Set the problem up for long division

3.5 ⟌ 8.4

3.5̣ ⟌ 8.4̇   No decimals in the Divisor!

```
      2.4
35 ⟌ 84.0
     70
     ‾‾‾
     140
```

8.4 ÷ 3.5 = 2.4

Framing Square Math

Converting Decimal Inches to Fractions of an Inch

Your square or tape measure is divided into ⅛ths, or ¹⁄₁₆th inches. Usually on the job, you need to manipulate the resulting measurements with a calculator. You're going to need to convert those measurements into "DECIMAL INCHES," and then convert them back again. The square can do all of that for you easily.

Otherwise, you can use a mathematical process (just multiply and divide). Which method you choose depends on TOLERANCE. How accurate of a measurement conversion do you need? We will assume a measurement of about 2.6", and do both methods.

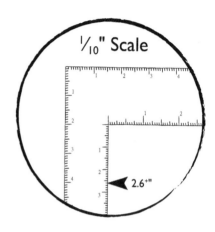

Fig. 1

You'll Need:
- A block of wood
- Your notebook (for the "Math" way)
- A Steel Square with the ¹⁄₁₀th scale

The "Square Way"
- Using the ¹⁄₁₀th scale on the BACK of the square, place your fingernail on the edge of the square at 2.6"
- Flip the square over.
- Read the measurement in ⅛" on the face of the square at your fingernail.
- It should be very close to 2 ⅝"

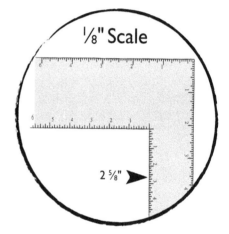

Fig. 2

The "Mathematical Way"
In "Let's do the Numbers" to the right, we'll convert 2.6 to inches and fractions of an inch.

To convert from decimal inches to fractional inches:
- Separate the WHOLE NUMBER from the fraction and set it aside. We'll use the remaining decimal, ".6."
- Decide what tolerance of measurement you need. In this case we'll work to ¹⁄₁₆ths.
- Multiply the remaining decimal by 16.
- The result is 9.6
- ⁹·⁶⁄₁₆ths is almost ¹⁰⁄₁₆, or ⅝
- 2.6 converts to roughly 2 ⅝"

Let's do the Numbers...
$2.6 = 2 + .6$
$.6 \times 16 = 9.6$
$^{9.6}/_{16} \approx ⅝$
$2.6 \approx 2 ⅝$

Or, set up as a ratio
(and using Algebra—
but don't tell anyone...)
$^{6}/_{10} = ^{x}/_{16}$
cross multiply
$6 \times 16 = 96$
Solve by Division
$^{96}/_{10} = 9.6$

Framing Square Math

Scale Drawing Using the ¹⁄₁₂" Scale

Twelve inches make a foot. That's why the ¹⁄₁₂" scales run along the outside edges of the **BACK** of a Framing Square. The ¹⁄₁₂" scale creates **RATIOS**: ¹⁄₁₂": 1" and 1": 1'. This enables the Framing Square to manipulate large measurements in "feet and inches" for calculation and layout.

In this exercise, you'll use the ¹⁄₁₂" scale on a Framing Square to make a 1":1' scale drawing of a 4'-5" by 7'-3" **RECTANGLE**.

Remember, the scale divides inches into ¹⁄₁₂ths, every inch on the square represents a foot in real life, and every ¹⁄₁₂" represents an inch.

You'll Need:
- Your notebook
- Framing Square (with ¹⁄₁₂" scale)
- Pencil

To Start:
- Find the **BACK** of the square. The outside edges should be marked in ¹⁄₁₂" scales. If it's not, find another square...
- On the paper, draw a right angle at least 5" by 8" (Fig. 1).
- On the 5" leg, measure 4 ⁵⁄₁₂" from the corner and make a mark (Fig. 2).
- On the 8" leg, measure 7 ³⁄₁₂" from the corner and make a mark (Fig. 2).
- Rotate the square to start completing the rectangle (Fig. 3).
- Line up the existing marks to the 4 ⁵⁄₁₂" and 7 ³⁄₁₂" measurements on the square (Fig. 3).
- Draw the lines to complete the rectangle (Fig. 3).

Use the Framing Square to make a 1":1' scale drawing of your classroom or shop. (You'll also need a tape measure.)

Fig. 1
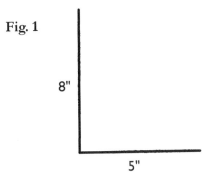

Fig. 2
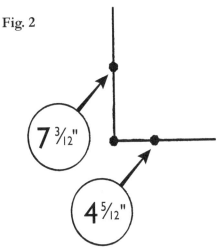

Fig. 3
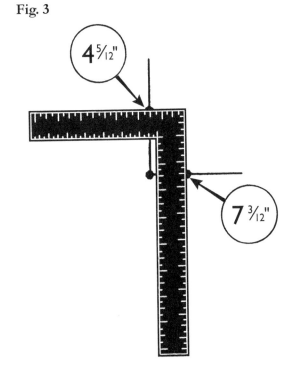

Calculating with Ratios and Proportions

As we saw in the last exercise, RATIOS can turn the Framing Square into an architect's scale ruler; but ratios and the Framing Square can do much more together. Ratios describe the relationship of one number, usually a measurement, to another. Equivalent ratios are called PROPORTIONS. Proportions allow you to solve for unknown answers. That's one of the major reasons we do math.

The Framing Square's ability to create SIMILAR TRIANGLES (Fig. 1) and the resulting proportions turn the tool into a calculator. To find an unknown measurement, you can set up the square with a known ratio and manipulate the square to create similar triangles that, in turn, create equivalent ratios (proportions.) The unknown measurement, will be contained in that proportion.

The Framing Square also allows us to calculate directly in feet and inches, or in inches and fractions of an inch. This is the way we measure. There's no need to convert to decimals, or IMPROPER FRACTIONS, in order to calculate, as there is with most electronic calculators. Avoiding an extra step avoids potential mistakes.

The Framing Square's ability to calculate accurately depends on your ability to measure, draw, and manipulate accurately. This type of calculation still depends on accuracy and organization, it just doesn't involve rows of numbers. Lots of calculations can be made with ratios and proportions. Ratios are simply relationships between things described in the language of math. We've just forgotten that many problems can be set up using ratios and proportions. The answers may not be "completely exact"; but for most measuring and building, they are certainly "close enough." This defines tolerance. Why calculate to ten decimal places when you can only measure to two decimal places?

The ratios presented in the following exercises have been compiled over many years to perform different tasks on the job.

We're going to do what old time carpenters called "SLIPPING THE SQUARE." You can manipulate the square with your hands and see the calculation happen in front of your eyes.

Let's do the Numbers...

Usually, math expresses ratios as fractions...

"Two to three" is writen as $\frac{2}{3}$ or 2:3

Proportions are equivalent ratios: $\frac{2}{3} :: \frac{6}{9}$

"Two is to Three as ? (What) is to Nine"

$$\frac{2}{3} = \frac{?}{9}$$

$\frac{2}{3} = \frac{?}{9}$ Cross Multiplication is used to solve for the unknown.

$$\frac{2 \times 9}{3} = ?$$

$$\frac{18}{3} = ?$$

$$6 = ?$$

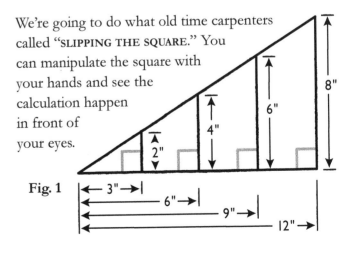

Fig. 1

Framing Square Math

Slipping the Square

SLIPPING THE SQUARE makes the tool a calculator. With this technique, you have to draw and measure accurately. If the square moves during your calculation, or you read a measurement incorrectly, your calculation will be incorrect.

Two Critical Components for Slipping the Square

- *A surface with a straight edge* – You need a surface that has a straight edge. You'll be reading measurements off the square against this edge. This can be a piece of plywood, or even a piece of cardstock or paper.
- *The Slip Line* – This line defines the ratio and resulting proportions that you will be computing. The square has to stay up against this line. Make sure you draw this line so you always have it as a reference.

Determined by the measurements of your desired, original, ratio, the angle of your square to the straightedge of your surface establishes the critical "slip line." By slipping the square along that line, you're creating larger, or smaller similar right triangles that are proportional to the original ratio. The results are read where the square intersects the straight edge of the surface.

You can set a straightedge, like a wooden ruler, along the slip line and slide the framing square against that straightedge. This works well when you have a partner.

You also might consider building a sliding bevel to use with the square for the next few exercises (see page 77, Making a Large Sliding Bevel). If you are working alone, this tool will tremendously increase the accuracy and speed of your calculations.

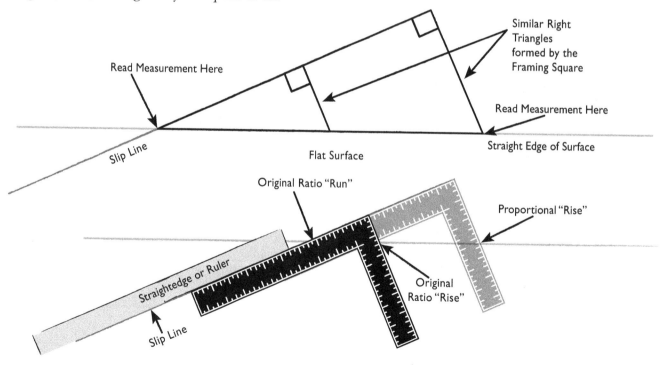

Framing Square Math

Multiplying with a Framing Square

Measurements get multiplied all the time on the job. Earlier, you multiplied mixed numbers and calculated the "traditional" way, with rows and columns of numbers. This time, the calculations will be done with SIMILAR TRIANGLES. Let's start SLIPPING THE SQUARE and multiplying. This takes practice. Be patient with yourself.

Let's multiply by 4. This uses the ratio of 1:4. It all works because of similar right triangles and "SLOPE." (Don't worry, we'll get there...)

Draw the "Slip Line" so that if your straight edge moves, you can reset it to the line. It also helps to have a partner hold the straightedge for you while you slip the square against it.

You'll Need:
- Framing Square
- Ruler/straightedge
- The edge of a surface (or larger piece of paper)
- Pencil

To Start:
- Set one leg of the square at 1" against a straight edge of paper (Fig.1).
- Set the other leg to 4" along the same edge of paper.
- Draw the "slip line" (Fig. 1).
- Put a straightedge (like a ruler) alongside the edge of the square where you drew the "slip" line. (You're going to "slip" the square along this edge.) Have someone help by holding down your straight edge.
- Slip the square so that the edge that was originally set at the 1" is now set at 1 ½". Make sure your straight edge hasn't moved (Fig. 2).
- The other leg should read 6", which is 4 x 1 ½".

Try multiplying 2 ½" × 4"

As a challenge, use the ¹⁄₁₂" scale of the Square to multiply 3'-6" by 3.

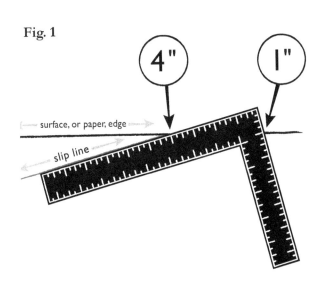

Fig. 1

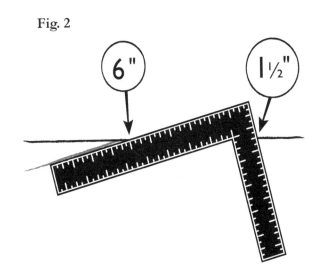

Fig. 2

Dividing with a Framing Square

Measurements also get divided on the job all the time. So, let's reverse the previous exercise and use the square to divide by four.

Fig. 1

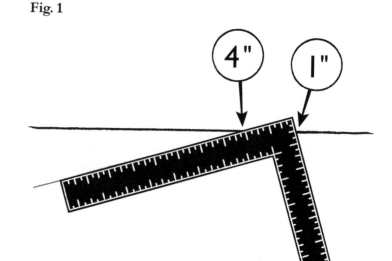

You'll Need:
- Framing Square
- Ruler/straightedge
- The edge of a surface (or larger piece of paper)
- Pencil

Let's see how to divide 8 by 4.

To Start:
- Use the same 1:4 "slipping line" we used in the previous exercise.
- Slip the square so that the 8" mark on the square's **BODY** lines up with the surface edge.
- What's the number on the square's **TONGUE** where it meets the surface edge?

Fig. 2

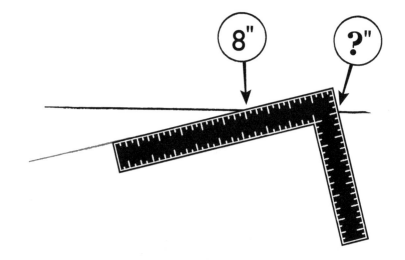

Try dividing:
6 by 4
10 by 4
3 by 4

As a challenge, try using the $^1/_{12}$" scale to divide 12'- 8" by 4

Framing Square Math

Why Does This Work?

The multiplication and division exercises we were just doing depend on establishing a ratio, changing one number and then determining the missing number of the proportion. All those similar right triangles we created within each exercise share the same vital ratio of one leg length to another. The relationship of the two shorter legs of a right triangle, or its "Rise over Run" is common in the building process and extremely useful (Fig. 2). Many of the exercises in this workbook depend on the concept.

The Rise over Run ratio (Fig. 2) is defined when you set the first measurements on the square against the edge of your paper, or board (Fig. 1). By **SLIPPING THE SQUARE** along that line, you're creating larger, or smaller similar right triangles that are proportional to one another.

Figures 3 and 4 show different ways to describe the relationship between triangles "A," "B," and "C."

RISE AND RUN is described everywhere in Math...

In Arithmetic ...A Fraction ...A Ratio
In **GEOMETRY** ...The **HYPOTENUSE** of a Right Triangle
In **ALGEBRA** ...The **SLOPE** of a line
In **TRIGONOMETRY** ...The **TANGENT FUNCTION**

No matter how you describe it, "Rise and Run" is a very powerful tool. Yet, it's all just the ratio of two legs of a right triangle.

Fig. 1
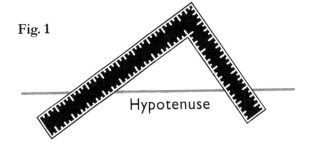

Fig. 2
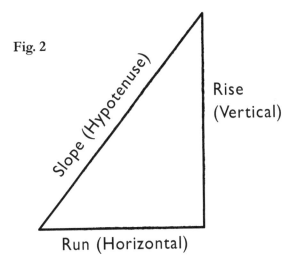

Fig. 3

$$\frac{Rise_A}{Run_A} :: \frac{Rise_B}{Run_B} :: \frac{Rise_C}{Run_C}$$

Fig. 4
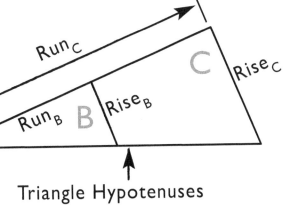

Triangle Hypotenuses

Framing Square Math

Finding Slope

On the job, you always have to determine unknown measurements. SLOPE is one of your most powerful tools. It can be used for small measurements, like the gap between drive shaft coupling faces on heavy machinery; or, the length of a roof rafter. Using the slope of a right triangle on the job depends on recognizing the "rise over run" in the "real life" situation. You can't use it if you can't see it.

You need to build your "eye."

We'll start off by drawing some specific triangles. Creating your own triangles with specific slope sharpens your skills (and your eyes).

Let's do an example. A triangle has a run of 4" and a rise of 3". Let's draw it.

You'll Need:
- Framing Square
- Pencil
- Paper

To Start:
- Place your square on the paper.
- Starting at the Square's VERTEX, draw a baseline 4".
- Make marks at both the Vertex and the 4" point.
- From the righthand end of that line, use the square to draw a 3" vertical line. Make a mark at the 3" point.
- Connect your marks with the triangle's HYPOTENUSE.

This hypotenuse has a slope of ¾

Use the square to draw the following triangles and determine their slope. The first number is rise, the second is run.

    2, 3
    6, 10
    3 ½, 14

Fig. 1

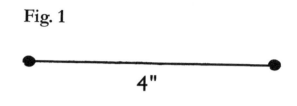

Fig. 2

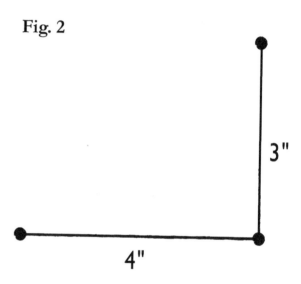

Fig. 3

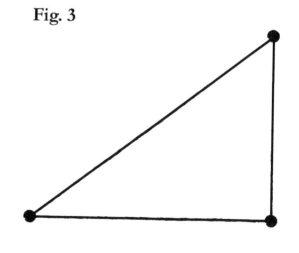

Using Slope to Find "Run" Lengths

Let's start using slope to figure out needed measurements. If we're in a situation where we know the slope of a line and we know our desired rise, we can figure out the needed run. This technique could be used to build a roof, or grade a ramp. As an example, let's use a ramp for wheelchair access. The ramp needs to have no more than a $\frac{1}{10}$ slope. The needed rise for the ramp is 2". How long is the ramp?

You'll Need:
- Framing Square
- Pencil
- Paper or drawing surface (at least 24" long)

To Start:
- On the paper, draw a long baseline and mark your "run" of 10" (Fig 1).
- Erect a 1" **PERPENDICULAR** from the 10" mark (Fig. 2).
- Draw the resulting triangle and extend the hypotenuse.
- Slip the square along your baseline until the extended hypotenuse crosses your desired rise: 2" (Fig. 3).
- Drop a line with the Square from that point on the slope to the baseline. Make a mark (Fig. 3).
- This is the distance of the run. It should be 20".

Now, you try a few examples.
Remember, slope = rise/run. Use the square to draw the following triangles. The slope is the first number. The rise is the second.
$\frac{2}{3}$, 3"
$\frac{1}{4}$, 1 $\frac{1}{2}$"
$\frac{4}{5}$, 2"

Follow the same process used in the example.
- *Draw the triangle of the desired slope.*
- *Extend your triangle's base line and hypotenuse.*
- *Slip your square along the baseline until the desired rise crosses the triangle's extented hypotenuse.*
- *Find the run.*

Fig. 1

Fig. 2
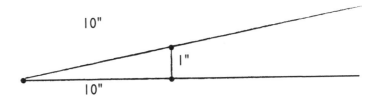

Fig. 3
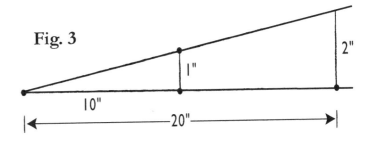

Let's do the Numbers...

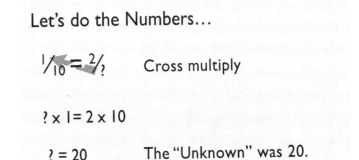

$\frac{1}{10} = \frac{2}{?}$ Cross multiply

? × 1 = 2 × 10

? = 20   The "Unknown" was 20.

Framing Square Math

Using Slope to Find "Rise" Lengths

Let's continue using slope to figure out needed measurements. If we're in a situation where we know the slope of a line and we know the intended run, we can figure out the needed rise. We'd use this on the job if we needed to frame a shed roof. Let's do an example where we know the slope of the roof is 1:3 and the distance between the supporting walls is 9'. (We'll use 1":1' scale.)

You'll Need:
- Framing Square
- Pencil
- Paper

To Start:
- Draw a baseline at least 10" long.
- From a starting point on the left of the line, measure 3". Make a mark (Fig. 1).
- Use the square and erect a 1" line from this point. Make a mark (Fig. 1).
- Connect the lines and extend the line out so that it extends as far as your baseline (Fig. 2).
- On the baseline, measure 9" from the starting point.
- Use the square and erect a line from this point until it intersects your extended "slope" line.
- Measure the line.
- This is your rise.

Use the square to draw the following triangles. The slope is the first number. The run is the second. Find the rise.

¾, 8
⅔, 12

Follow the same process used in the example.
- *Draw the triangle of the desired slope.*
- *Extend your triangle's base line and hypotenuse.*
- *Mark a point on the extended baseline.*
- *Slip your square along the baseline to this point.*
- *Find the rise.*

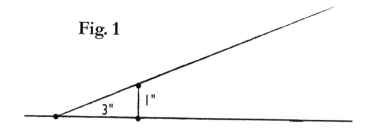

Fig. 1

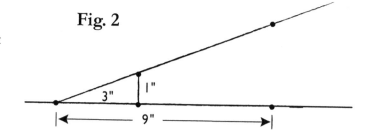

Fig. 2

Let's do the Numbers...

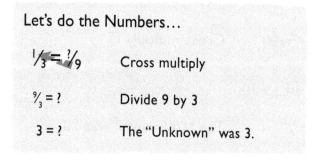

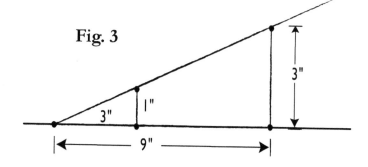

Fig. 3

Circles and Pi

Next to a straight line, circles are the simplest shape to draw. It only takes the most basic tools, a pair of dividers, or even just a piece of string and a stick. With those tools and a straightedge, you can also cut the circle into most of the shapes you use on the job. Circles, straight lines, 90 degree, 45 degree, 30 degree, and 60 degree angles are all easily created. The foundation of so much math, circles are also tremendously useful on the job. Laying out the foundations of a building, designing a truss system for a roof, or making a pattern for the leg of a dining room table can all be done with parts of a circle.

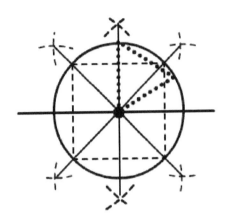

The two most important measurements of a circle are the distance around its outside (the **CIRCUMFERENCE**) and the distance across its middle (the **DIAMETER**). Basic human curiosity led someone a long time ago to see what happens when you divide one of these numbers by the other. The resulting ratio created by dividing a circle's circumference by its diameter is called "**Pi**" (π). The formula is
$\pi = C/D$.

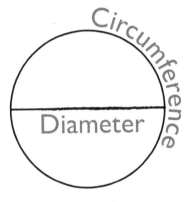

For a simple ratio, Pi is a complicated number. It's true, absolute, value can never be fully determined; it can only be approximated. (For mathematicians, this makes Pi an "irrational number.") The value of Pi is usually said as an approximate decimal 3.14159. Pi also can be approximated as a fraction ($^{22}/_7$) whose decimal value equals 3.14286. (That's a one hundred and twenty seven hundred thousandths difference between the fractional and decimal values—not much.)

We'll be using Pi's fractional approximation, $^{22}/_7$, because it's a ratio.

Framing Square Math

Finding Circumference

On the job, you usually know the circle's diameter and need to figure out the length of its circumference; or, most times, a part of that circumference. This happens when you have to put drywall or baseboard on a curved wall, or band the outside edge of a table. On a large scale, you may need to figure out the number of regularly spaced columns needed for a circular shaped building.

As we just saw, the distance around a circle (**CIRCUMFERENCE**) and the distance across the middle of a circle (**DIAMETER**) make a ratio called "**Pi**" (π). The formula is $\pi = C/D$. We'll use the approximate fractional value of Pi ($^{22}/_7$) to "slip the square" and find circumference.

You'll Need:
- Framing Square
- Ruler/straightedge
- The edge of a surface (or larger piece of paper)
- Pencil

To Start:
- Set the tongue of the square to 7" and the **BODY** to 22" (you can also use 11" and 3 ½"). The tongue is the diameter's measurement, the body the circumference's (Fig. 1).
- Line up the straightedge along one edge of the square (the slip line).
- Draw the slip line.
- Slip the square to answer the following questions:
    - *What's the circumference of a circle whose diameter is 2 ½"? (Fig. 2)*
    - *What's the circumference of a circle whose diameter is 1"? (Fig. 3)*

Fig. 1

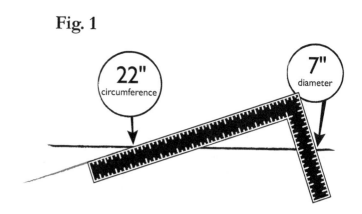

Fig. 2

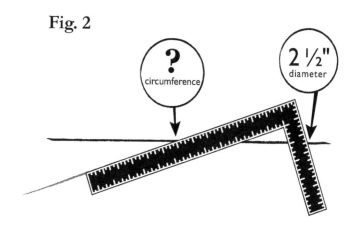

Fig. 3

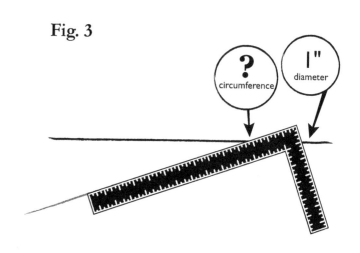

Ways to Figure the Area of Circles

Next to RECTANGLES, circles are the most commonly built shape. Buildings, windows, stairs and tables can all be shaped like circles. Many walls are built along a portion of a circle. It's important to be able to figure out the dimensions of existing, or desired circles. Finding the area of a circle tells us how much material we will need to build the shape and how much paint will be needed to cover it.

In a traditional math classroom, we learn the formula for the area of a circle is Pi times the radius squared.
$A = \pi r^2$. This formula requires a lot of multiplication.

```
    8.5          72.25
  × 8.5         × 3.14
  ─────         ──────
   4 2 5         28900
  6 8 0 0        72250
  ──────       2167500
  72.25        ───────
               226.8650
```

Luckily, there's a ratio we can use on the Framing Square!

It depends on an alternative version of the formula for a circle: $A = \tfrac{1}{4}\pi \times D^2$
(A = Area, D = Diameter)

The drawings below show how the radius calculates ¼ of the surrounding squares' area and the diameter calculates all of the surrounding sqaures' area. Therefore, in the formulas, the radius is paired with π and the diameter is paired with ¼ π.

A little math tells us that the ratio for a quarter π:
$^{22}/_7 \times ^1/_4 = ^{22}/_{28} = ^{11}/_{14}$

11/14 is our ratio. With it, all we have to do is "square" the diameter and slip our Framing Square. So, if you can figure the area of the surrounding square and remember ¼ π is approximately $^{11}/_{14}$, you can figure out the area of a circle.

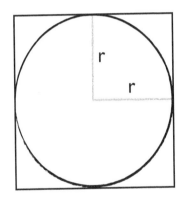

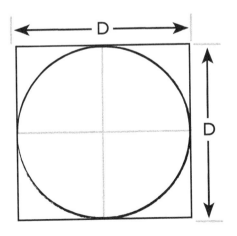

Framing Square Math

Calculating the Area of Circles

Now that we've covered all that theory, let's put it to work and find the area of a circle with a diameter of 3" (Fig. 1). Remember, an alternative version of the formula for a circle is: $A = \frac{1}{4}\pi \times D^2$ (A = Area, D = Diameter)

You'll Need:
- Framing Square
- Ruler/Straight edge
- The edge of a surface (or larger piece of paper)
- Pencil

To Start:
- We're going to use the ratio of 11 to 14 ($\frac{1}{4}\pi$).
- Set the tongue of the square to 5 ½" (11 halves) against a straight edge of paper (Fig. 2).
- Set the other leg to "7" (14 halves) along the same edge of paper (Fig. 2).
- Put a straightedge alongside the 7" edge of the square.

Our circle has a diameter of 3. Three squared is 9.

- Slip the square so that the edge that originally was set at 7 is now set at 9.

The other leg should read "7 1/16," which is the area in SQUARE MEASUREMENT (square inches) of our 3" diameter circle.

Now, find the areas of circles whose diameters are 2" and 4".

This ratio also works for determining the area of Ellipses.

ELLIPSES are perfect ovals. They're not as common on the job as circles, but they still show up a lot (Fig. 3). Some examples of ellipses on the job are transom windows over doors, movie screens (sometimes), and lots of details in furniture. The Area of an Ellipse equals ¼π times the area of the rectangle (base × height) described by that ellipse.
$A = \frac{1}{4}\pi \times (B \times H)$ (A= Area, B= Base, H= Height)
Remember this; it could be useful if you ever work with ellipses.

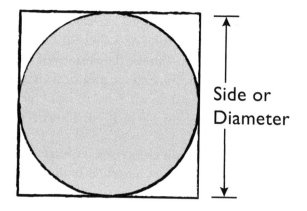

Fig. 1 — Side or Diameter

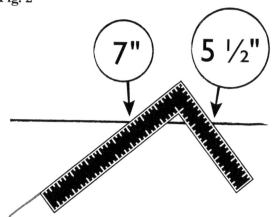

Fig. 2 — 7"   5 ½"

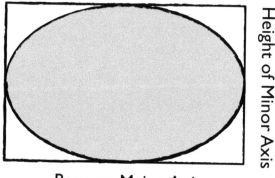

Fig. 3 — Base or Major Axis / Height of Minor Axis

Finding Squares and Circles with the Same Area

Sometimes you know the area of a circle and have to find the square that has the same area (Fig. 1). Water could be flowing from a round pipe to a square sectioned pipe. Maybe, you have a water tank that is being filled by round pipes; but the only way you have to drain it is by cutting a square hole?

The need for this calculation doesn't happen often on the job; but it happened often enough for this formula to be documented in several old Framing Square books. It's neat to be able to solve something this complex with a simple ratio of 85:96. "85" is the side of the square. "96" is the diameter of the circle.

Let's try an example where we have a 10" round pipe and need to cut a square hole with the equivalent area.

You'll Need:
- Framing Square
- Ruler/Straight edge
- The edge of a surface (or larger piece of paper)
- Pencil

To Start:
- Let's use "eighths" as our units. Whole inches are too big.
- Set the Framing Square to $^{85}/_8$ and $^{96}/_8$—10 $^5/_8$" and 12" (Fig. 2). Remember, 85 represents the side of the square, and 96 represents the diameter of the circle.
- Slip the edge representing the pipe (the one set to 12") down to 10".
- The other leg should read 8 $^7/_8$".

What is the size of the equivalent squares if the diameters of the circles are: 5" and 15" ?

What are the diameters of the circles if the sizes of the equivalent squares are 11 $^1/_2$" and 6" ?

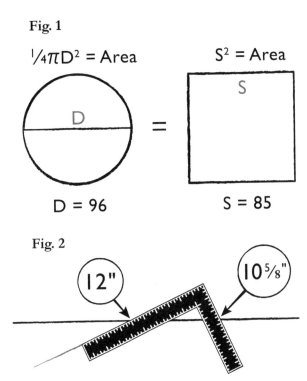

Fig. 1

$^1/_4 \pi D^2$ = Area S^2 = Area

D = 96   S = 85

Fig. 2

12"   10 $^5/_8$"

Why this Works...

$^1/_4 \pi \times D^2 = S^2$

$^1/_4 \times ^{22}/_7 = ^{22}/_{28} = ^{11}/_{14} = ^1/_4 \pi$ Assume D = 2
$\pi = ^{22}/_7$

$^{11}/_{14} \times (2)^2 = S^2$

$^{11}/_{14} \times 4 = S^2$

$^{44}/_{14} = ^{22}/_7$

$\sqrt{\dfrac{22}{7}} = S$ If the Diameter is 2, the side is the square root of π.

$\sqrt{\pi} \cong 1.77$

$\sqrt{\dfrac{\pi}{2}} \cong .885$

$\dfrac{85}{96} \cong .885$

$\dfrac{85}{96}$ (Side of the square)
       (Diameter of the circle)

Dimensioning Proportional Rectangles

Proportional rectangles often get drawn during the building process. Floor layers make proportional inlays. Furniture makers make proportional drawer faces; and architects design proportional windows. We'll use the $\frac{1}{12}$ inch scale on the outside edge of the back of the Framing Square to draw a 3'-6" by 6'-7" rectangle in 1":1' scale.

You'll Need:
- Framing Square
- Pencil
- Paper

To Start:
- Find the 12ths scale on the back face of the Framing Square. Remember, $\frac{1}{12}$":1" in 1":1' scale
- On a piece of paper, draw a right angle at least 6" wide by 9" tall (Fig.1).
- On the 6" base, measure and mark 3" and $\frac{6}{12}$ths from the corner (Fig. 2).
- On the 9" vertical leg, measure and mark 6" and $\frac{7}{12}$ths from the corner (Fig. 2).
- Draw the DIAGONAL connecting the two marks from the top left, to the bottom right creating a right triangle (Fig. 2).

Let's find the proportional rectangle whose width will be 2' in 1":1' scale.
- Turn the square so that the tongue (short leg) is facing to the right and the body (long leg) is facing up.
- Slip the square to the right until the tongue reads 2" (Fig. 3).
- What is the reading on the square's body? This is the length of the proportional rectangle.

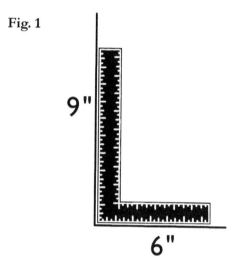

Fig. 1

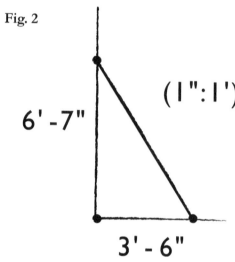

Fig. 2

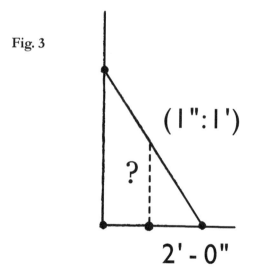

Fig. 3

Making an Octagon with 7/24ths

Octagons often occur in buildings. Columns, windows, and the walls of a building can all be formed in the shape of an octagon.

Fig. 1

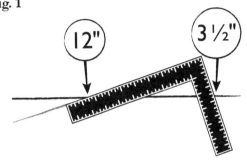

The General Process:
- Determine the size of the square inside which you will draw your octagon.
- Divide one side of the square into 24 equal parts (Fig. 2).
- Measure seven of those parts in from each corner of the square (Fig. 2).
- Connect those points to form the octagon (Fig. 2).

If you're creating a large octagon, you can use the 1/12ths scale on the back of the square to calculate the distances in feet and inches. Let's layout an octagon inside of a 6" sided square.

Fig. 2

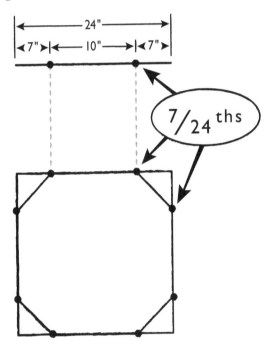

You'll Need:
- Framing Square
- Ruler/Straightedge
- The edge of a surface (or larger piece of paper)
- Pencil

To Start:
First, let's use the Framing Square to calculate how far in from the surrounding square's corner we need to measure (Fig. 1).
- Set the square to the 24:7 ratio. The ratio of 12" on the body to 3 ½" on the tongue works well.
- Slip the square to 6" on the body.
- The measurement on the tongue should be 1 ¾"
- Write it in your notebook. This is the distance in from the corners of the square you will need to measure in order to outline your octagon.

Now, let's draw the octagon.
- Draw a 6" square.
- Mark the calculated distance in from the corners (1 ¾").
- Draw the resulting octagon...

Draw the octagons that are formed within squares having the following side lengths: 8", 12 ½", 14 ⅜"

Let's do the Numbers...

$7/24 \times 6 = ?$ Rewrite the problem as an improper fraction.

$$\frac{7 \times 6}{24} = ?$$

$$\frac{42}{24} = 1\ ^{18}/_{24} = 1\ ^{3}/_{4}$$

Framing Square Math

Pythagorean Theorem
$a^2+b^2=c^2$

Pythagoras of Samos, an ancient Greek who lived around 570 BC, was the focus of philosophic, religious, and mathematical movements. These movements revolved around the many cool things that happen with, or can be described by, right triangles.

Those ideas have influenced western thought ever since.

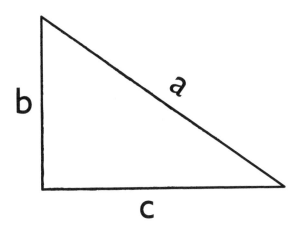

The areas of the squares formed by the two shorter sides of a right triangle equal the area of the square formed by the long side (HYPOTENUSE) of that triangle:

$$a^2 = b^2 + c^2$$

Sometimes this equation solves with WHOLE NUMBERS. These are called "Pythagorean triples." The most common triple is:

3, 4, 5
(shown above)

Here are a few other Pythagorean Triples that describe different shaped triangles:

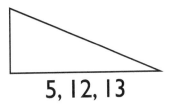
5, 12, 13

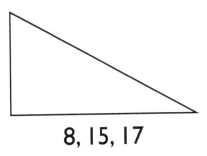
8, 15, 17

7, 24, 25

Framing Square Math

Proving the Pythagorean Theorem

Let's prove the **Pythagorean Theorem**! (Don't get too excited.)

Using the framing square we can draw the illustration on the right for a 3, 4, 5 triangle. The areas of the squares made from the triangle's sides prove the theorem.

You'll Need:

- Framing Square
- Pencil
- Paper

To Start:

- Use the square to draw a 3" by 4" by 5" right triangle.
- Extend right angles from the corners.
- Create 3", 4", and 5" squares from the appropriate sides.
- Divide each square into 1" squares.
- Count the resulting squares.

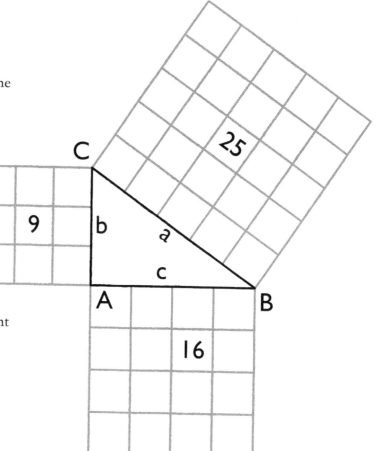

Knowing how this tool works is very useful when you need to create a right angle. Or, if you need to check that an existing angle truly is 90 degrees.

Different triples create differently shaped right triangles. Since not all spaces in a building, or pieces of furniture, fit a 3, 4, 5 triangle, these differently shaped right triangles can be very useful.

Prove another triple—say 5", 12", 13". You'll need a larger piece of paper, or drawing surface.

Finding the Diameter of a Circle Whose Area is Equal to the Area of Two Given Circles

Two different-sized pipes empty water into a tank. What size pipe is needed to drain the tank? Determining the answer depends on the area of the cross sections of the pipe. The problem can be solved with the Framing Square by finding the diameter of a circle whose area is equal to the area of two given circles.

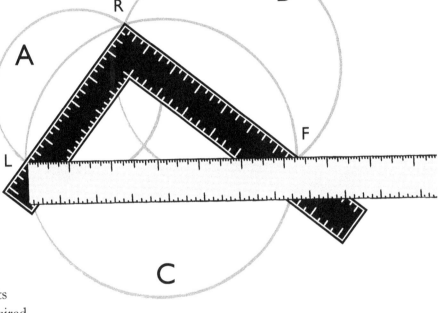

You'll Need:
- Framing Square
- Pencil
- Paper

To Start:
- On the square's tongue, mark the diameter (RL) of one of the given circles (A).
- On the square's body, mark the diameter (RF) of the second given circles (B).
- The distance between the points (LF) is the diameter of the required circle (C).

Why This Works...We Know:
- The "alternative" theorem to πr^2 for the area of a circle is, $\frac{1}{4}\pi \times D^2$
- The **Pythagorean Theorem** is $a^2+b^2=c^2$
- And this exercise says that the area of circle A plus the area of circle B equals the area of circle C.
- We're going to write that mathematical sentence combining the formulas and labeling the diameters with "subscript" "a"s, "b"s and "c"s: $\frac{1}{4}\pi \times D_a^2 + \frac{1}{4}\pi \times D_b^2 = \frac{1}{4}\pi \times D_c^2$

- Cancel the $\frac{1}{4}\pi$ from each side of the equation, we end up with:
$D_a^2 + D_b^2 = D_c^2$

...which is the Pythagorean Theorem and the reason why this exercise works.

Try finding the diameter of a circle that combines the areas of 5" and 17" diameter circles.

Framing Square Math

Layout Exercises

Remember, Framing Squares were (and still are) used to build stuff. Big stuff, medium-sized stuff, and small stuff.

Small layout happens within the dimensions of the square.

In medium-sized layout, lines continue beyond the dimensions of the square.

In large layout the square projects square lines and makes calculations.

The tool truly allows us to do "rare," or refined, layout, as that old Framing Square book mentioned in the introduction says.

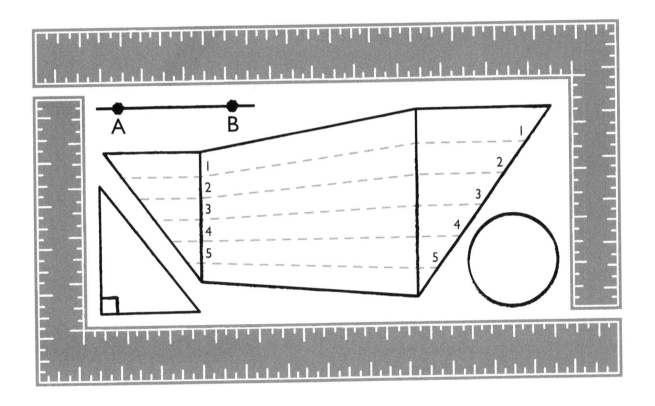

Framing Square Math

Dividing a Line into Equal Parts

Measurements need to be divided all the time. Columns need to be spaced; railings and windows need to be sized and placed.

We've used traditional math and slipping the square to do division. Now, we're going to use the square to create the correct right triangle and then subdivide that triangle with evenly spaced, similar right triangles.

You'll Need:
- Framing Square
- Pencil
- Paper
- A surface into which you can nail (optional)
- Finish nails (optional)
- Hammer (optional)

To Start:
- Draw the line that needs to be divided. Make sure it is shorter than the long leg of the Framing Square. This will become the hypotenuse of the right triangle (Fig. 1).
- Let's divide this line into nine parts.
- Place the end of the line at the 9" mark on the body of the square (driving in a finish nails helps) (Fig. 2).
- Rotate the square so that the other end of the line just touches the inside of the short leg (tongue) of the square (Fig. 3).
- Draw the inside of the square to complete the triangle.
- Next, mark the inch lines along the edge of the body—number them 1 through 9 (Fig. 4).
- From these points, use the square to erect vertical perpendicular lines that intersect your original line—the one we're trying to divide (Fig. 4).
- The intersections mark the divisions of the original line.

Now, draw a line that fits on your paper and use this technique to divide that line into seven equal parts.

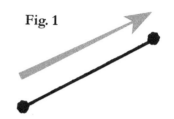

Fig. 1

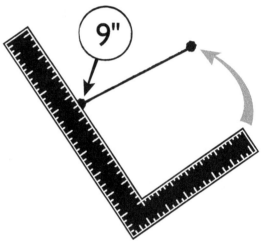

Fig. 2

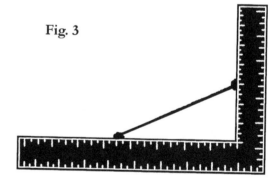

Fig. 3

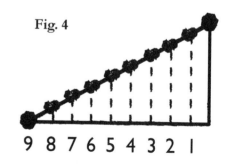

Fig. 4

Framing Square Math

Dividing a Board into Equal Parts

We're going to divide a board into five equal parts using the Framing Square to create a **DIAGONAL** line that touches both edges of the board. The diagonal line needs to be a length easily divisible by 5.

This exercise shows up in lots of old carpentry books. It uses similar triangles the same way as the previous exercise. This is a skill which is still useful on the job.

You'll Need:
- Pencil
- A length of dimensioned lumber 8" or wider, at least 2' long (you can also use a piece of 8 ½" x 11" paper)
- Framing Square

To Start:
- Select a measurement greater than the width of the board and easily divisible into five parts.
- For example, for an 8" wide board, lay the square so that 0" (the **VERTEX**) is at one edge and 10" is at the other edge of the board.
- Draw a diagonal line along the edge of the square.
- Mark points on the board at 2", 4", 6", and 8".
- Repeat the process at the other end of the board and connect the points with horizontal lines.
- The space has now been divided into 5 equal parts.

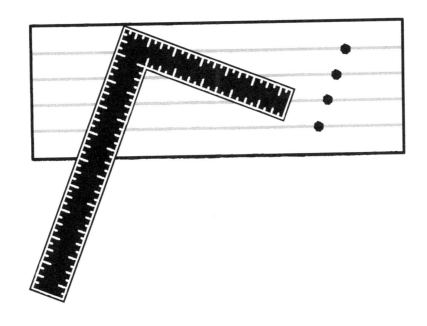

Framing Square Math

Laying Out Polygons Within a Circle

Here's another useful, "old school," exercise that always seems to work; but won't be found in any math textbook. This method can be used to lay out polygons with any number of equal sides—squares, pentagons, octagons, you name it. It builds on several techniques we've already used.

We'll draw a pentagon (a 5-sided polygon) as an example.

Please note: For clarity in these illustrations, the Framing Square is drawn smaller than described in the text.

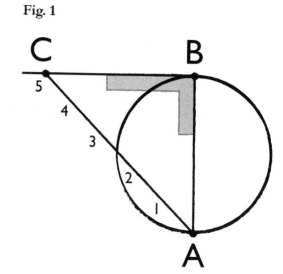

Fig. 1

You'll Need:
- Framing Square
- Drawing surface
- Nails or pins
- Pencil
- Compass

To Start:
- Use the 16" square's tongue as the diameter.
- Draw the diameter AB.
- Drive nails at points A and B. Then, draw a circle with the square using the technique described earlier on page 9. You'll need to flip the square to complete the circle.
- Use a square to draw a line perpendicular from an end of the diameter (point B). Extend, or project, that line BC. Don't worry about its specific length, yet (Fig. 1).
- From point A draw a line that intersects line BC at a distance easily divided by 5 (Fig. 1).
- Mark the equally spaced points on line AC and label them 1–4.
- Slide the square along line AB with its body pointing towards the diagonal line AC. Mark the points along lines AB when the line extended from the square's body intersects the "1 through 5" points on the diagonal line AC. These points should be labeled 1 through 5, as in the illustration (Fig. 2).

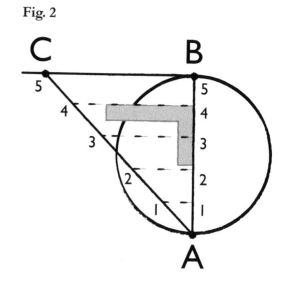

Fig. 2

54   Framing Square Math

- From Points A and B use the body of the square to swing 16" arcs to the right of the circle. Hook the inside corner of the square on the nails at A and B. Put your pencil at 16" on the inside edge of the body, and swing the arcs. The arcs should cross outside the circle at Point D (Fig. 3).
- From Point D, draw a line through point "2" on the diameter. The line needs to extend far enough to cross the circumference of the circle past the diameter and create Point E (Fig. 4).
- The distance EA is one face of the desired Pentagon.
- Use dividers, or a compass, to measure that face.
- Walk the dividers around the circumference of the circle and mark the resulting points (Fig. 5).
- Finish laying out the pentagon by connecting those points (Fig. 5).

This technique works for polygons with any number of faces. So remember, if you:
- Establish the size of the circle that will surround your polygon.
- Divide the circle's diameter by the desired number of faces in your polygon.
- Establish a point outside circle by swinging arcs the length of the diameter from the endpoints of the diameter.
- Draw a line through that point through Point Number 2 on the diameter so that it crosses the far side of the circumference.

You will determine the length of one face of the polygon.

Try creating 6- and 8-sided polygons.

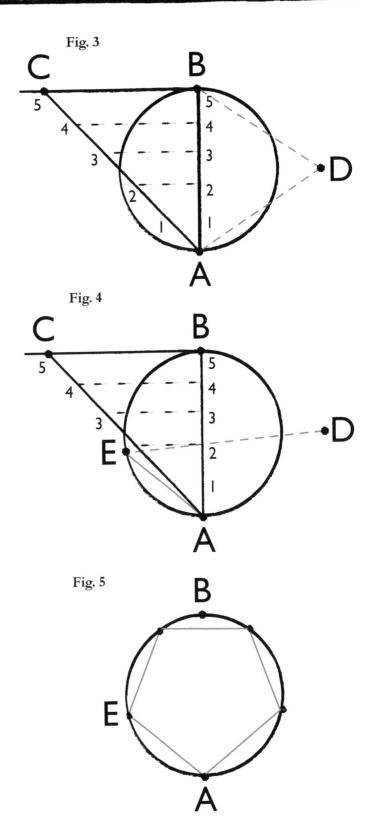

Fig. 3

Fig. 4

Fig. 5

Framing Square Math

Dividing a Longer Line into Equal Parts

In this exercise a line still needs to be divided; but it is too long for the necessary similar right triangles to be laid out inside the Framing Square. So, the Framing Square is used to produce a larger set of similar triangles.

You'll need:
- Pencil
- Framing Square
- Tape measure
- Chalk line
- Chalk (optional)
- Masking Tape (optional)

To Start:
- You'll use your chalk line to make all the long lines. Mark your points with chalk or pencil. For "temporary" layout, you can use chalk, or make your pencil marks on pieces of masking tape.
- Lay out the line, which you'll divide into nine parts. We'll call it line AB.
- Lay out a second line from point A that's easily divided into nine equal parts. (We'll call this line AC.)
- Label the resulting eight marks "a" through "k."
- Connect points B and C with a line (BC).
- From any point on BC, use a square to draw a line perpendicular to line BC that extends almost past point A. This is line DE.
- Slide the square along line DE with its body pointing towards line AC. Mark the points along DE when the line extended from the square's body intersect the "a through k" points on line AC.
- Extend these lines so that they intersect your original line AB (if they haven't crossed it already). These intersections will divide line AC into eight equal parts.

See any familiar shapes?

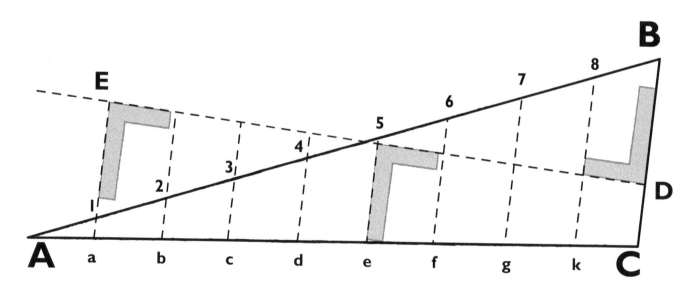

56 Framing Square Math

Dividing a Tapered Space into Equal Parts

Sometimes you need to divide a space that isn't a rectangle into equal parts. Here's a way to do it. As in the previous exercise, a framing square extends the perpendicular lines.

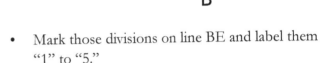

You'll Need:
- Pencil (you can also use chalk, if you need to make the layout "temporary")
- Framing Square
- Chalk line
- Tape measure

To Start:
- Given the size of your drawing surface, create an **IRREGULAR** space ABCD similar to the illustration. We're going to use similar triangles to divide lines CA and DB into six equal spaces.
- Draw line CF 90° to line CA and to its left.
- Draw line DE 90° to line DB and to its right.
- From Point A, draw a diagonal line (AF) to the left of line AC, so that it intersects line CF. The length of AF should be easily divided into six parts.
- Mark those divisions on line AF and label them "1" to "5."
- From Point B, draw a diagonal line (BE) to the right of line BD so that it intersects DE. BE should be a length easily divided into six parts.
- Mark those divisions on line BE and label them "1" to "5."
- Slide the square along lines AC and BD with its body pointing towards the appropriate diagonal line. Mark the points along lines AC and BD when the lines extended from the square's body intersect the "1 through 5" points on the diagonal lines AF and BE. You can use a chalk line to extend the lines, as needed.
- Label these points 1 through 5, as in the illustration, and connect the "1 through 5" lines from line AC to those on line BD. The irregular space ABCD is now divided into six equal parts.

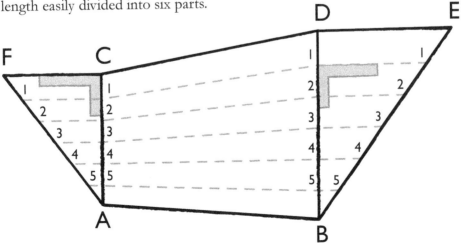

Framing Square Math

Laying Out a 10 Degree Angle

Another "Old School" technique contained in almost every book about Framing Squares concerns laying out specific **ANGLES**. Given measurements on the Framing Square's tongue and body a desired angle can be laid out. Architects and designers make small angles on small drawings using a **PROTRACTOR**. Full size angle layout isn't so convenient; although it's not hard. We'll use a framing square and the provided Angle Table to draw a 10 degree angle. This technique uses "**RISE AND RUN**"; and we've seen them before. So, don't be intimidated.

You'll Need:
- Drawing surface – paper, plywood, etc.
- Framing Square
- Pencil

To Start:
- Orient your drawing surface so you have at least 22 inches horizontally and 6 inches vertically.
- Draw a horizontal line at least 20" long.
- Make a point at the far right hand side of the line.
- Lay the square so the body faces to the left and the tongue faces up.
- Using the Angle Table for Square, look up the Tongue and Body measurements for creating a 10 degree angle. (They are: Body 19.7", Tongue 3.47".)

These measurements are in decimal inches. To use the scales on the framing square, you have two choices:
1. Convert the decimals to 16ths of an inch. In order to do so, multiply the decimal of an inch by 16. After rounding, this will yield the number of 16ths in the decimal.
2. Use the ¹⁄₁₀" scale on the square. To do this, use any scale to mark the whole number of the "body" measurement on your line. Then, add the missing decimal of an inch with the ¹⁄₁₀" scale.

Once you've decided on your technique and have converted the decimal measurements from the table into 16ths, or are ready to use the ¹⁄₁₀" scale, see the illustration and:

- Lay the body of the square on the line and position the inside scale so that 19.7" is at the point on the line you have marked.
- Draw a vertical line along the inside of the tongue.
- Using the table, mark a point the correct tongue height along the vertical line.
- Use the square as a straightedge to connect the vertical and horizontal points.

The resulting line should be very close to 10 degrees. You can check the angle with a protractor.

Try laying out:
18° and 30° angles.

Angle Table For Square								
Angle	Tongue	Body	Angle	Tongue	Body	Angle	Tongue	Body
1	0.35	20	16	5.51	19.23	31	10.28	17.14
2	0.7	19.99	17	5.85	19.13	32	10.6	16.96
3	1.05	19.97	18	6.58	19.02	33	10.89	16.77
4	1.4	19.95	19	6.51	18.91	34	11.18	16.58
5	1.74	19.92	20	6.84	18.79	35	11.47	16.38
6	2.09	19.89	21	7.17	18.67	36	11.76	16.18
7	2.44	19.85	22	7.49	18.54	37	12.04	15.98
8	2.78	19.81	23	7.8	18.4	38	12.31	15.76
9	3.13	19.75	24	8.13	18.27	39	12.59	15.54
10	3.47	19.7	25	8.45	18.13	40	12.87	15.32
11	3.82	19.63	26	8.77	17.98	41	13.12	15.09
12	4.16	19.56	27	9.08	17.82	42	13.38	14.89
13	4.5	19.49	28	9.39	17.66	43	13.64	14.63
14	4.84	19.41	29	9.7	17.49	44	13.89	14.39
15	5.18	19.31	30	10	17.32	45	14.14	14.14

Framing Square Math

Drawing an Ellipse—Locating the Foci

ELLIPSES are another shape that gets used a lot on the job—everything from sidewalks to movie screens contain portions of ellipses. Framing squares alone can't really draw ellipses; but a square will figure out the measurements you need to draw an ellipse using the "pin and string" method (see p. 60). The Framing Square will also help you layout the grid on which you'll draw the ellipse.

The standard way to locate the foci of an ellipse is to swing an arc the length of half of the **MAJOR AXIS** (line AB in Fig. 4) from the top of the **MINOR AXIS** (line CD in Fig. 4). The two points where this arc crosses the Major Axis are the **FOCI** (points F and G in Fig. 4). You can also locate the foci by drawing a triangle with the Framing Square. (If you were laying out a large ellipse, you would be able to do this in "feet and inches" using the ¹⁄₁₂ths scale on the back of the Square.)

You'll Need:
- Pencil
- Framing Square
- Paper

To Start:
- Draw a line that is half the length of the Ellipse's major axis (Fig. 1).
- Locate the square so that half of the ellipse's minor axis is the distance from the inside of the Square's vertex to the left-hand point of the line you just drew (Fig 2).
- Rotate the square until the inside of the tongue touches the other end of the ½ Major Axis line (Fig. 3).
- Draw the Triangle.
- The distance from the inside vertex of the Square to the point you just located on the inside of the Square's tongue is the distance from the Ellipse's **ORIGIN** (Fig. 4, Point E) to the Foci (Fig. 4, Points F and G).

You can also draw an ellipse by combining the Framing Square with a tool you can make, the **TRAMMEL**. See the exercise on page 78.

Fig. 1

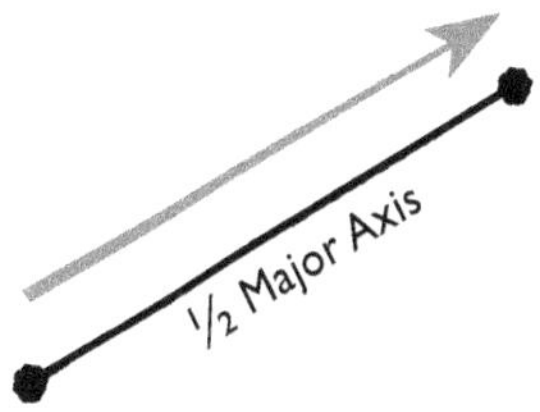

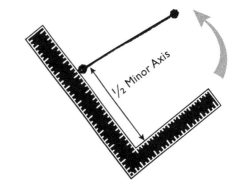

Fig. 2

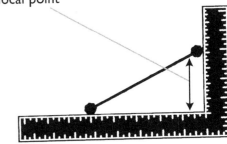

Fig. 3

Distance from the center (origin) to focal point

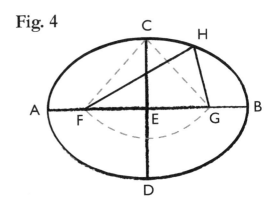

Fig. 4

Framing Square Math

Drawing an Ellipse—Pin and String Method

ELLIPSES are perfect ovals. They are described by the height of their MINOR AXIS (Line CD in Fig. 1) and the length of their MAJOR AXIS (Line AB in Fig. 1). Your Major and Minor axis are also x and y axis. (You're drawing on a COORDINATE PLANE.)

The challenges to drawing the desired sized ellipse with this technique are to locate the foci and accurately cut the string to the length of the major axis. If you want to draw a larger ellipse, you can use nails instead of pins and stranded wire instead of string. (Wire is better than string because it doesn't stretch as easily.)

If you know the intended size of your ellipse, you already know the length of the major axis and the length of the minor axis. However, in order to use this "Pin and String" method to draw an ellipse, you need to locate the foci (FOCAL POINTS) on the ellipse's Major Axis. After you find that distance (see the previous exercise), you can locate your "pins" and use the "string" to draw your ellipse.

You'll Need:
- Drawing surface something into which you can push a pin or drive a finish nail
- Hammer
- Pins or finish nails
- String or stranded wire
- Pencil/Chalk
- Straightedge/chalk line
- Framing Square
- Compass
- Paper

Drawing the Major and Minor Axis of the Ellipse
- Layout the Coordinate Plane, as in "Creating the Coordinate Plane," page 63. This will be lines AB and CD in Fig. 1.

Mark the Foci on the Major Axis
- Set your compass to the distance of line AE, half the MAJOR AXIS.
- Put the metal point of the compass at point C. Swing two arcs that intersect the Major Axis at points F and G. These are your Focal Points, or Foci. Or, you can use the Framing Square based technique in the previous exercise.
- Measure the focal distance from Point E and mark the Foci (Points F and G) on the Major Axis (Line AB).
- Insert the pins, or nails, at the foci.
- Make a piece of string, or wire, that has loops on its ends and is the length of the major axis.
- Loop the ends of string over the pins.
- Put a pencil on the inside of the string and hold it so that the string is taught.
- Draw the ellipse making sure to keep the string taut.

Fig. 1

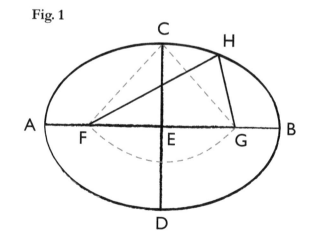

Framing Square Math

Coordinate Planes and Algebra

The right angles and scales contained in Framing Squares can create the Coordinate Plane that describes much of basic Algebra and Statistics.

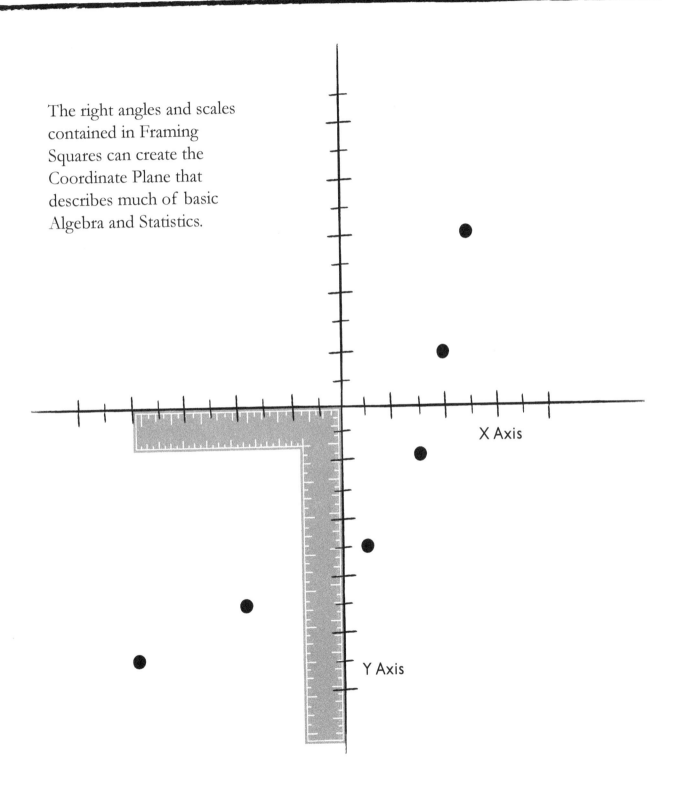

Framing Square Math

Descartes and Galileo

Knowing a little of the history of something makes it more interesting, and helps make it more applicable in our world. Math is no different.

Renee Descartes, a brilliant French mathematician, scientist, and philosopher, lived in the early 1600s. These were dangerous times for an innovative thinker who said, "I think, therefore I am."

Galileo, born thirty years earlier, had been persecuted for his mathematical and scientific beliefs. Descartes took notice and stayed on the road, avoiding persecution, for most of his life.

Descartes

For us, as users of practical math, Descartes conceived of negative numbers and the **COORDINATE PLANE**. Galileo invented the sector. The sector is another tool that is a mechanical calculator, like the Framing Square. It also depends on similar triangles, The sector's calculations and measurements are read with dividers. On the job, the two-foot long folding rule would be used as a sector. The folding rule and dividers are central to the Carpenters' Union emblem. It's not an accident.

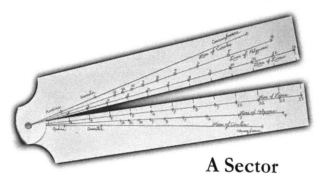
A Sector

Galileo

Creating the Coordinate Plane and Coordinates

Let's draw and label Descartes' COORDINATE PLANE.

You'll Need:
- Framing Square
- Drawing surface/paper (at least 14" by 14")
- Pencil

To Start:
- Mark a point of beginning (ORIGIN) in the center of the paper.
- Place the vertex of the square at the origin. Have the tongue facing up and the body facing to the right (Fig. 1).
- Draw the outside "vertical" and "horizontal" edges of the square (Fig. 2).
- Mark and number every whole inch. Put a "+" sign in front of each number (Fig. 2). Mark the "half inches" as well.
- Using the body of the square as a straight edge, extend the vertical below the point of origin (Fig. 2).
- Place the vertex of the square at the origin. Have the tongue facing to the left and the body facing down.
- Align the body along the line you just drew. Make sure the corner of the square is still on the "origin" (Fig. 3).
- Draw the outside edges of the square.
- Mark and number every whole inch. Put a "-"sign in front of each number (Fig. 4). Mark the "half inches" as well.

You've just created the coordinate plane, a great format for laying out and positioning any shape. The horizontal line is the "X" axis. The vertical line is the "Y axis." The POSITIVE AND NEGATIVE NUMBERS are your "X and Y Coordinates."

When you write the coordinates of a point as pairs of numbers, the "X" coordinate is the first number; the "Y" coordinate is the second. In Fig. 4, the point (-2, 3) is "3 up" from the origin and "two to the left" of the origin.

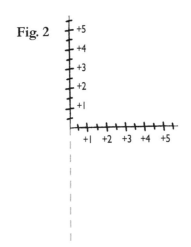

Fig. 2

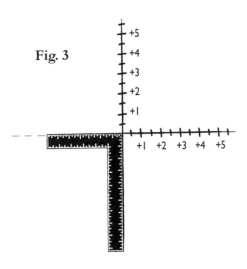

Fig. 3

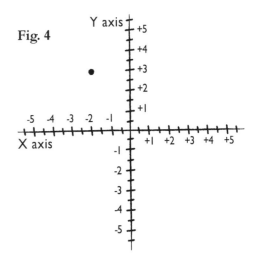

Fig. 4

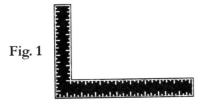

Fig. 1

Framing Square Math

Locating Points on the Coordinate Plane

When you're laying out a project, you have to start from somewhere. This is your **Point of Beginning**, or **Origin**. From this point draw your first line. On this "control line," you lay out points from which other lines extend.

On big jobs, this work can be done with a wide variety of tools; but it all comes back to what we're going to do now. This concept is critical to siting structures on properties, as well as laying out buildings and rooms within those buildings.

Let's lay out the foundation of a house.

Using the coordinate plane you've created in the previous exercise, locate and label the following points that describe the foundation of a house. (Remember, the "X" coordinate is the first number; the "Y" coordinate is the second number.)

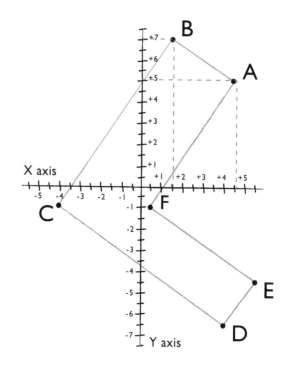

A (+4 ½, +5)
B (+1 ½, +7)
C (-4, -1)
D (+4, -6 ½)
E (+5 ½, -4 ½)
F (+½, -1)

Use your square to connect the points as shown and measure the distances between the points:

AB
BC
CD

Use the coordinates to measure the distances between C and F.

How could we measure the distance between A and B using their coordinates? Hint: Pythagoras...

"Distance" is a great example of using "Absolute Value." It doesn't matter if the points are located in the positive, or negative areas of the coordinate plane. The distance between them is still a distance traveled—a positive value.

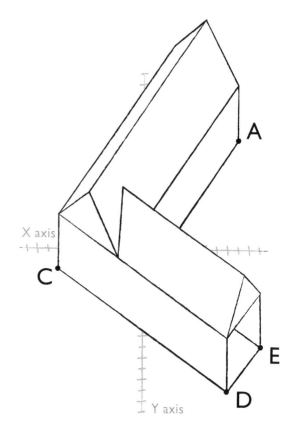

Framing Square Math

Checking Layout on the Coordinate Plane

Let's continue our layout project...

We've drawn a good looking picture; but are our corners 90 degrees? What's it going to be like building this house? Let's start finding out the quality of our layout by using our square to double check to see if we have right ANGLEs.

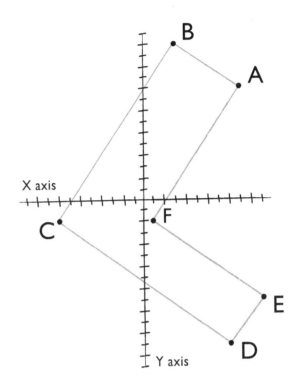

You'll Need:
- Your Framing Square
- Your drawing

To Start:
- Use your square to check if the following ANGLEs are greater than 90 degrees, less than 90 degrees, or exactly 90 degrees.
- Fill out the chart below with your results.

	>90	<90	90
<ABC			
<BCD			
<CDE			
<DEF			
<EFA			
<FAB			

Framing Square Math

Finding Slope with the Framing Square

On the job, the ability to layout, or check, PARALLEL and PERPENDICULAR lines is critical. If a building's foundation is out of square, the effects of that poor layout will be felt throughout the whole building process.

Rise/run = slope

Slope tells us a lot about our lines and their relationship to one another, which is very helpful when doing large scale, or very precise, layout.

- Lines are parallel if they have the same slope. A line with a slope of 8/5 is parallel to a line with a slope of 32/20 (they are equivalent fractions).
- Perpendicular lines have slopes that are the negative INVERSE of each other. (This means that the ratio is inverted and the sign changes.) A line with a slope of 8/5 is perpendicular to a line with a slope of -5/8.

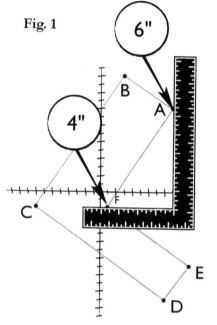

When figuring the slope of a line on the coordinate plane, you work from left to right.

- If the line goes "uphill" the line has a positive slope (+).
- If the line goes "downhill" the line has a negative slope (-).

Let's use the Square to find the Slope of Line FA by measurement.

You'll Need:
- Your drawing from the previous exercise
- Framing Square
- Pencil
- Notebook

To Start:
- Place the square "body up" on the drawing, so it touches points A and F (Fig.1).
- Align the tongue parallel with the X axis.
- Measure the "Rise" and "Run" of the triangle that has the "measured" line (AF) as its hypotenuse.
- In your notebook, write the measurement of the "Rise" over the measurement of the "run." Remember to simplify.
- The slope of line AF is 6/4, or simplified 3/2. Since it goes "uphill," the slope is positive, +3/2.

Fig. 2

Line	Slope
CB	
FA	
DE	
CD	
FE	
BA	

Fig. 3

Line	Parallel to Line?	(Y/N)	Perpendicular to Line	(Y/N)
CB	FA		CD	
FA	DE		FE	
CD	FE			
FE	BA			

Use your Square to find the Slope of the following lines: CB, BA, CD, DE
- Fill in the chart (Fig. 2) with each line's slope.
- Remember to indicate if the slope is "+" or "-"

Figure out:
- Which lines are truly 90 degrees (perpendicular)?
- Which lines are parallel?
- Fill out the chart in Fig. 3

Finding Slope with Algebra

We've seen how important **Rise and Run** are when building things. It's also critical to how **Algebra** describes lines. Algebra defines slope as, "The change in the Y value over (divided by) the change in the X value."

The Formula is: $\dfrac{y_2 - y}{x_2 - x} = \text{slope}$

This formula just numerically calculates the distances between the points' horizontal and vertical measurements—the same thing you did with the Framing Square in the last exercise.

Continuing to use our drawing, let's find the slope of lines FA and CB using the points' coordinates, and the Algebraic formula. Don't worry, it's still just rise over run. (And, it's all written out in the "Let's do the numbers…")

You'll need:
- Pencil
- Your Notebook
- The Carpenters Math Checklist (page 80) is also very helpful.

To Start:
- Write the Algebraic formula for Slope.
- Write the coordinates for points F and A.
- Plug the numbers into the formula. F will be our first point; A will be our second point.
- 6/4 is our Rise over Run. Simplified, it equals 3/2
- Repeat the process for Line CB
- How do these numbers compare to the slopes you measured in the previous exercise? Are the lines truly **parallel**?
- Algebra gives us a more exact number than measuring a triangle created with the square. The Algebraic method applies if we should find ourselves needing to build to a high tolerance.

If you want more practice,
- *Go back to the Locating Points on the Coordinate Plane exercise (page 64) for the coordinates of points.*
- *Determine the lines' slopes Algebraically and fill out the chart in Fig. 2 of the previous exercise.*
- *Determine which lines are parallel and perpendicular.*
- *Fill out the chart in Fig. 3 of the previous exercise.*

Let's do the numbers...
Are lines CB and FA Parallel?

Find the slope of line FA with coordinates:

F = (½, -1)
A = (+4 ½, +5)

$\dfrac{y_2 - y}{x_2 - x} = \text{slope}$ Figure out the distances of the "rise" and "run"

$\dfrac{5 - (-1)}{4½ - ½} = \dfrac{5 + 1}{4} = \dfrac{6}{4} = \dfrac{3}{2}$

slope FA = 3/2

Find the slope of the line CB with coordinates:

C = (-4, -1)
B = (+1½, +7)

$\dfrac{y_2 - y}{x_2 - x} = \text{slope}$ Figure out the distances of the "rise" and "run"

$\dfrac{7 - (-1)}{1½ - (-4)} = \dfrac{7 + 1}{1½ + 4} = \dfrac{8}{5½} = \dfrac{8}{11/2} = 8/1 \times 2/11 = 16/11$

slope CB = 16/11

$16/11 \neq 3/2$

The Lines CB and FA are not parallel!

Framing Square Math

Linear Equations

Big projects, like highways and office buildings, depend upon accurate measurement of long lines. Many times those lines need to be projected over long distances; or, they need to be continued after being interrupted by other structures. Algebra and the Coordinate Plane enable us to do this type of work.

If we know the equation of a line when it happens in a part of a coordinate plane, we can project that line to other parts of that coordinate plane. And, we can locate points on that line anywhere in the space described by our coordinate plane.

As we said, most Algebra can be described on the coordinate plane. An Algebraic equation of any straight line is called a LINEAR EQUATION. "Linear" just means straight. A standard form of the equation for a straight line is:

$y = mx + b$

x ...is the X coordinate of a specific point
y ...is the Y coordinate of that same point
m ...is the slope of the line
b ...is the y-intercept of the line

The y-intercept is just the point where the line crosses the y-axis.

As an example, let's figure out the equation of line CB. Then let's figure out the Y coordinate of a point on that line, if the X coordinate is +5.

Follow the process outlined to the right in your notebook.

You'll Need:
- Your notebook
- Pencil

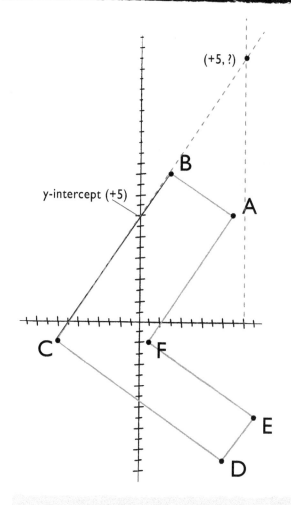

Let's do the Numbers...

$y = mx + b$	Write the formula for a line.
slope CB = $^{16}/_{11}$	Determine the slope of the line CB (we know this from the previous excercise).
$b = 5$	Determine "b," the y-intercept from the drawing
$y = ^{16}/_{11} x + 5$	Construct a formula for line CB.
$y = (^{16}/_{11} \times 5) + 5$	Now, let's extend the line CB. to find the y-coordinate of a point that has an x-coordinate of 5.
$y = ^{80}/_{11} + 5$	
$y = 7\,^{3}/_{11} + 5 = 12\,^{3}/_{11}$	The "y" coordinate is a little more than 12 ¼ for an "x" coordinate of 5.

Framing Square Trigonometry

On the job, sometimes you need to calculate lengths; sometimes you need to calculate angles. Measurement over long distances and very precise measurement are made much easier by **TRIGONOMETRY**. All the word "Trigonometry" means is the "measurement of triangles"; something we've been doing throughout this book. "Trig" is a tremendously useful tool, and nothing to be scared of. Learning how to use it is just a "next step" in becoming a better builder.

In basic Trig, there are three triangles that need to be measured. They all can be drawn and measured with a Framing Square and come out of the same "Unit Circle" (See Fig. 1).

Unit means "one." When we divide any number by one, it stays the same. So, when we're working with the Trig **FUNCTION**s, if the bottom number of the ratio is one, the value of the Trig function is the ratio's top number. Using these triangles, this number is a measurement.

The "Trig Functions" are just ratios that use the lengths of these triangles' sides (Fig. 2). Each angle has its own set of ratios. If you know the angle and one leg of the triangle, you can find the lengths of the other two legs. Or, if you know two legs of a right triangle, you can find the angle. A few things to remember when "Talking Trig":

- The "**ADJACENT**" side is always the short leg next to the angle we are measuring.
- The angle we're measuring is often called "Theta." It's another Greek letter, θ.
- The Trig functions are ratios. (The **TANGENT** function is just "Rise over Run.")

Two acronyms that can help you remember the Trig formulas are in Fig. 3.

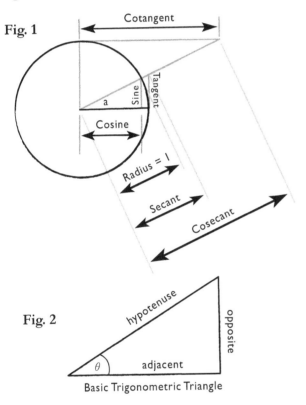

Fig. 1

Fig. 2 — Basic Trigonometric Triangle

Trigonometric Functions Formulas...	
Opposite side/Hypotenuse	= SINE
Adjacent side/Hypotenuse	= COSINE
Opposite side/Adjacent side	= TANGENT
Adjacent side/Opposite side	= COTANGENT
Hypotenuse/Adjacent side	= SECANT
Hypotenuse/Opposite side	= COSECANT

Fig. 3

Useful Trigonometric Acronyms...

Oscar   Opposite/Hypotenuse = SINE
Has

A   Adjacent/Hypotenuse = COSINE
House

Of   Opposite/Adjacent = TANGENT
Apples

S — Sine / Opposite
O
H — Hypotenuse
C — Cosine
A — Adjacent
H — Hypotenuse
T — Tangent
O — Opposite
A — Adjacent

Framing Square Math

Drawing the Trigonometric Functions
(Sine and Cosine)

Let's draw the first of the three Trig triangles. The **Sine** and **Cosine** Triangle

You'll need:
- Pencil
- Paper
- Framing Square

To Start:
- Use the square to draw a right angle with horizontal, and vertical, 10" legs. These are both radii of our Unit Circle. We're using 10" as our unit (Fig. 1).
- From the 10" mark on the horizontal leg, raise a 7 ½" perpendicular and make a mark (Fig. 2).
- Draw a line from the vertex of the right angle through this point (Fig. 2). This creates our angle, which we'll call θ (the Greek letter "Theta"). This line is also the hypotenuse of our first Trig triangle.
- Measure 10" from the vertex along the hypotenuse (Fig. 3).
- Make a mark (Fig. 3).
- Raise a perpendicular line from the baseline to this point (Fig. 3).

We've now created the triangle that contains our angle's **SINE** and **COSINE** (Fig. 3). And, we can measure them.

- Using the ¹⁄₁₀" scale on your square, measure the lengths of sine and cosine lines.
- The Sine function is the length of the line opposite our angle θ divided by the Hypotenuse (which is our unit, 10").
- The Cosine is the length of the line **ADJACENT** to our angle θ divided by the Hypotenuse (which is our unit, 10").
- Since our unit is "ten," not "one," we have to divide by 10. Just move our measurement's decimal place one space to the left. For example, if you measure 6" for your Sine value, the actual sine value is .6 (rounded to the nearest hundredth). Look it up on a Trig table on page 71.

A Sine of .6 belongs to a 37 degree **ANGLE**.

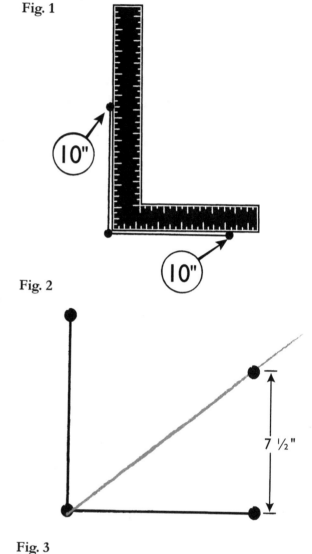

Fig. 1

Fig. 2

Fig. 3

Framing Square Math

Drawing the Trigonometric Functions
(Tangent, Secant, Cotangent, and Cosecant)

Let's continue adding to our Trig drawing by creating the **Tangent** and **Secant** Triangle,

You'll need:
- Pencil
- Paper
- Framing Square
- Your previous Trig drawing

To Start:
Use the square to raise a perpendicular line from the 10" mark on the baseline so that it crosses the "angle line." This vertical line is the Tangent.
- The hypotenuse of the triangle is the Secant
- Using the $\frac{1}{10}$" scale on your square, measure the values of secant and tangent.
- Move the decimal point one place to the left on your measurement.
- Look up the Tangent value on a Trig table (or on an electronic calculator, computer, etc.).

It should belong to an angle very close to the angle you determined using the **sine** and **cosine** functions.

To draw the **Cotangent** and **Cosecant** Triangle
- Use the square to create a right angle from the 10" mark on the vertical line. Extend it until it crosses the angle line. You may need to extend the angle line.
- Measure the Cotangent and the Cosecant.

Many math teachers can't tell their students when they'll ever use a Cotangent or Cosecant. All they need to do is ask a Millwright. The ability to calculate those distances without having to reset sophisticated measuring equipment can save significant time and money on a job.

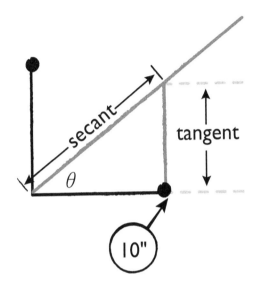

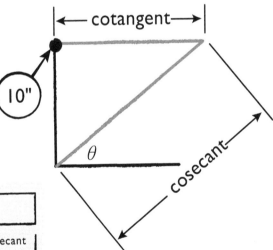

Trigonometric Table for Angles 35, 36, 37, 38...						
Angle°	Sine	Cosine	Tangent	Cotangent	Secant	Cosecant
35	.5736	.8192	.7002	1.428	1.221	1.743
36	.5878	.8090	.7265	1.376	1.236	1.701
37	.6018	.7986	.7536	1.327	1.252	1.662
38	.6157	.7880	.7813	1.280	1.269	1.624

Framing Square Math

Finding Length Using Tangent

If we know two facts about a right triangle, Trig allow us to find the lengths of its legs, or its angles. This is amazingly useful throughout the building industry. Surveying land, laying out rafter lengths in roofs, precision machinery alignment, and even GPS depend on Trig! Let's use Trig to find the lengths of some lines.

Using the Tangent Function

Just another example of the most popular ratio in the building process, rise over run, the Tangent function is the most used Trig function in the Building Trades. We'll use a simplified roof truss as our example. (Fig. 1) The roof has a 10:12 slope; meaning that the triangle that makes half of the roof has 10 units of rise to 12 units of run. A 10:12 slope roof has approximately a 40 degree angle at its base. (Fig. 2)

You'll Need:
- Pencil
- Your notebook

To Start:
- Work the problem outlined on this page in your notebook.

Then, figure a problem where the roof has the same pitch, the joist is 22' long and you need to know the height of the Center Post. Remember to use the Carpenter's Checklist on p. 80.

If the joist that runs along the base of our triangle is 16' long, how high is the center post?

Our Angle = 40°
Adjacent side = 16'
(Joist length)

What do we know?

Opposite side = ?
(Height of center post)

What do we want to know?

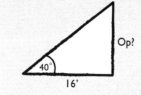

Draw a picture.

$$\text{Tan} = \frac{\text{Op}}{\text{Adj}}$$

Determine the correct formula

Tan 40° = .839

Look up Tangent function for 40°

.839 = ?/16'

Write the equation.

80% of 16?

Make an estimate, do some "mental math."

75% (¾) of 16' would be 12'

So, the answer would be a little more than that ...about 13'

.839 × 16' = 13.424'

Calculate.

13.424' = 13' 5 1/16"

Check against your estimate.

This is the height of your Center Post.

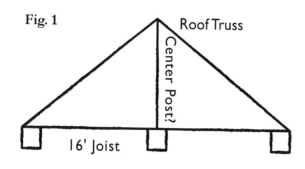

Fig. 1 — Roof Truss, Center Post?, 16' Joist

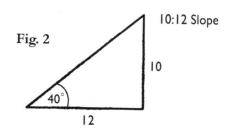

Fig. 2 — 10:12 Slope, 10, 12, 40°

Finding Length Using Cosine

Using the Cosine Function to find rafter length

Given the same roof truss as on the previous page, and all you've learned, use the COSINE function to find the rafter length.

You'll Need:
- Pencil
- Your notebook

To Start:
- Work the problem outlined here in your notebook.

Then, figure a problem where the roof has the same pitch, the joist is 22' long and you need to know the length of the rafter. Remember to use the Carpenter's Checklist on p. 80.

Fig. 1

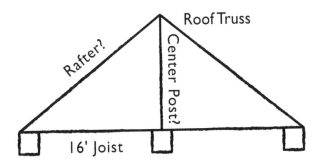

Fig. 2

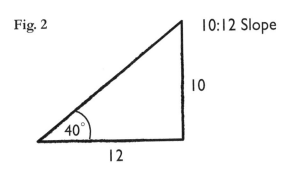

If the joist that runs along the base of our triangle is 16' long, how long is the rafter?

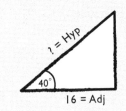

Draw the picture.

∠ = 40°   We know the angle and the adjacent side.

Adjacent side = 16'   We need to find the length of the hypotenuse.

$$\text{Cosine} = \frac{\text{Adj}}{\text{Hyp}}$$ What is the correct formula?

Cos 40° = .766   What is the value of the Cosine of 40°?

.766 = ¹⁶'/?   Plug numbers into the formula.

? = ¹⁶'/.766

.766 ≈ ¾

16' is ¾ of what number?   Estimate.

...about 20

16' ÷ .766 = 20.89'   Calculate.

The rafter length is:

20.89' = 20'-10 ⅝"   Check against your estimate.

Measuring 10 Degrees Using the Tangent Function

This exercise builds on a previous exercise, Laying Out a 10 Degree Angle (page 58). In this case, we're just going to use the back of the square, its ¹⁄₁₀ths scale and our Trig tables. It's actually so easy, you'll wonder why anyone gets intimidated by Trig.

You'll Need:
- Your Square
- Paper
- Pencil
- A Trig table (paper or electronic) that gives you the value of the tangent of 10 degrees

To Start:
- Place your square on the paper so the back face of the square is up and you have enough room to draw a 10" line along the inside edge of the body.
- From the inner vertex of the square, draw a 10" line along the inside edge of the body and mark 10".
- Draw a vertical line from the inner vertex of the square along the inside edge of the tongue.
- Look up the value of the tangent of 10 degrees (.176).
- Multiply it by 10" (the length of our base line) ...you'll get 1.76".
- Measure that distance up your vertical line, using the 10ths scale and make a mark.
- Connect this mark with your horizontal 10" mark.

You just created a 10 degree angle. You can mark 10" on the horizontal, erect a perpendicular and then use the tangent table and the same process to create any angle.

That's a powerful tool!

Go ahead and lay out 15, 25, and 30 degree angles.
Tan 15° = .268
Tan 25° = .466
Tan 30° = .577

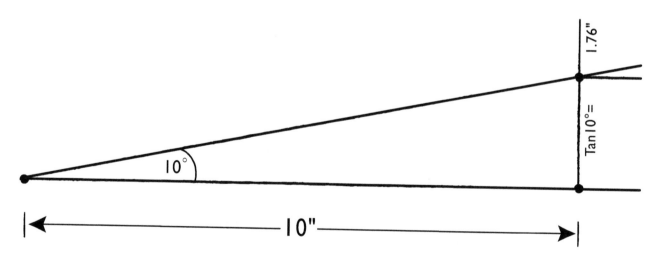

Framing Square Math

Build Your Own Tools

Throughout history, builders have built their own tools. Many times, they had no choice. There were no other tool builders nearby. Other times, they could build a better tool for their job than they could buy. Both those situations still apply today. Here are some tools you can build to help you do some of these exercises—and sometimes do them faster, and better.

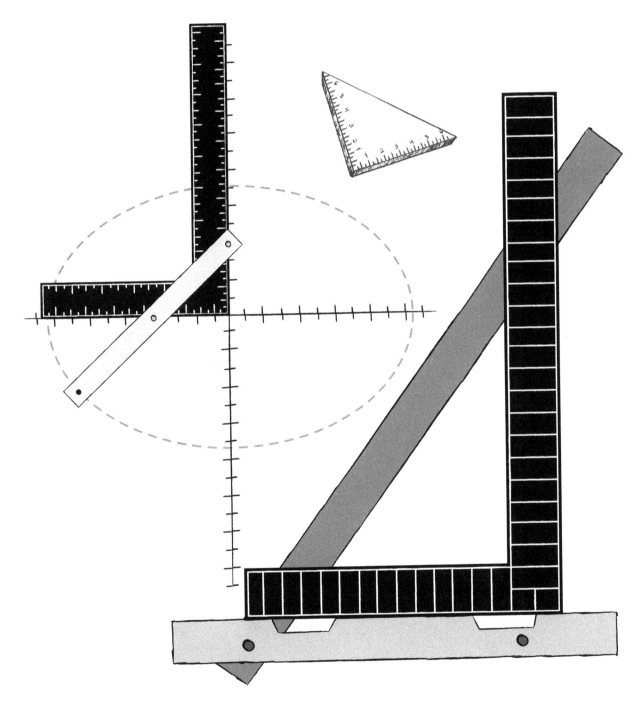

Framing Square Math

Making Your Own Framing Square

One of the joys of being a "builder" is making your own tools. You can make a simple Framing Square out of cardboard, or cardstock. It won't have all the different scales and tables that are on a purchased Framing Square; and it's "inches" won't match any other ruler; but it will do most of the calculations presented in this book. And, it will be YOURS.

You'll Need:
- Cardboard – at least 8 ½" x 11". "Shirt cardboard" or heavy cardstock work well.
- Scissors
- Compass
- Pencil

To Start:
- Either, find a piece of cardboard with a right angle corner, or draw and cut out a right angle on your piece of cardboard (Fig. 1).
- Set a compass to a small distance—less than the width of your pinkie finger (Fig. 2).
- Starting from the 90 degree corner, mark both edges using the compass to set the distance between marks. You want about 25 marks on each edge. The marks shouldn't go too deep into the piece of cardboard—about half the distance between marks (Fig. 3).
- Starting with the second mark from the 90 degree corner, make every other mark twice the depth of the original mark (Fig. 4).
- Now, make every fourth mark 3 times the depth of your original marks (Fig. 5).
- At this point you can label the deepest marks "1," "2," "3," etc. This will give you a ruler with wholes, halves and quarters (Fig. 5).

Fig. 1

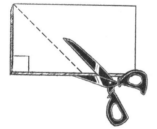

Fig. 2

Fig. 3

Fig. 4

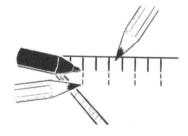

Fig. 5

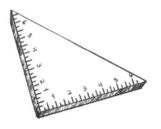

Making a Large Sliding Bevel

We've seen how "slipping" turns a square into a geometric calculator. A large "BEVEL GAUGE" makes slipping the square much easier. If there was one tool to build to complement the Framing Square, this is it. A hundred years ago, there were some bevels commercially available. Not now. Luckily, the tool is easy to build.

You'll Need:
- Two ¼" x 1 ¼" pieces of wood with straight edges for the legs (Base 20" long, Blade 26" long).
- ¼" x 1" Carriage Bolt with washer and wing nut
- Ruler or Square
- Drill with a ¼" drill bit
- Coping saw or band saw

(Large paint stirring sticks will make a smaller bevel with 17" by 17" legs. It works; but you won't be able to make as wide a range of calculations.)

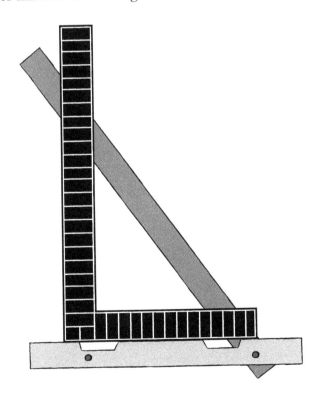

To Build the Base:
- Measure 3 ½" in from each end of the 20" long piece. Use a square to draw a line across the face of the base at both marks.
- Mark ⅜" up from the bottom edge of the base on both lines.
- Drill a ¼" hole at both of these points for the carriage bolt.
- To lay out the notches, measure 2 ½" in from each end and make a mark.
- From each mark, continue towards the center of the base 3 ½" and make another set of marks.
- Draw ¼" deep notches between the two sets of marks.
- Cut the notches with a coping saw, or a band saw.

To Build the Blade:
- Measure 1" from an end of the 26" long piece. Use a square to draw a line across the face.
- Drill a ¼" hole, on center along this line.

Assembly:
Decide whether you'll be "slipping" using the face, or back, of the square. (Which scales will you need?) If using the face, the blade will be carriage bolted on the bevel's left end. If using the back, the blade will be carriage bolted on the bevel's right end. Have the wing nut and washer facing towards you. The blade should lie under the square.

Use:
The blade of the Sliding bevel replaces the "straight edge of the surface" in our ratio exercises. The square now "slips" along the base of the Sliding Bevel.

Using a Wooden Trammel To Draw an Ellipse

Builders from furniture makers to drywall installers many times have to repeat the same sized ellipse. A wooden TRAMMEL makes drawing a specific sized ellipse easy. The Framing Square becomes a guide along which the trammel rides. A pencil held by the trammel draws ¼ of the specific ellipse, which serves as the pattern for the remaining three quarters of the shape. Let's draw an ellipse with a MAJOR AXIS of 8" and a MINOR AXIS of 5 ½".

You'll Need:

- A drawing surface (8 ½" x 11" paper is fine)
- Wood – ¼" x 1 ½" x longer than half of the ellipse's major axis. In our case, 6" works fine.
- Double-sided tape
- Wooden dowel (³⁄₁₆") two pieces ¾" long
- Pencil
- Framing Square
- Awl
- Drill (maybe a drill press)
- ³⁄₁₆" drill bit for the dowels
- ¼" drill bit for the pencil hole (usually)

To Start:

- On your drafting paper, select a center point "C." Draw the major axis of the ellipse, horizontally through C. The distance from point A to point C should be half of the length of the major axis (Fig. 1).
- Erect a line perpendicular to the major axis from the center point C. Mark half the length of the minor axis on this line and label the line CB.
- Complete the minor axis by extending the line an equal length below point C.
- **On the 6" piece of wood**, draw a centerline down the middle of the 1 ½" face. All marking will be done on this centerline (Fig. 2).
- Starting ½" in from one end, mark point "a." From "a," measure the distance AC. Mark point "c" at this distance on the trammel. From "c"

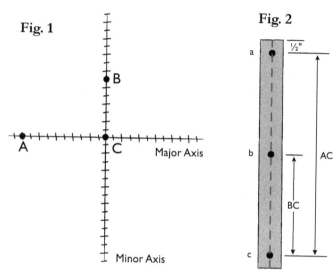

measure the distance BC back towards point "a" and mark point "b" on the trammel.
- Use your awl to mark points "a, b, and c" for drilling.
- Drill ³⁄₁₆" holes at points "a" and "b."
- Drill a hole (usually ¼") to snugly hold the pencil at point "c."

The trammel will draw the quarter of the ellipse on the opposite side of the major axis from where the Framing Square is placed (Fig. 3).

To Draw the Ellipse:

- Place the square so that its outside corner is on the center of the ellipse and the body and tongue of the square are aligned with the ellipse's major and minor axis (Fig. 3).
- Put two small strips of double stick tape on the back of the square to temporarily hold it on the drawing.
- Run the dowels of the trammel along the outside edges of the square and the tool will draw a quarter of your ellipse.

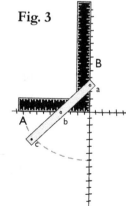

Framing Square Math

Appendices

Useful Tips

Outdoor Projects For Groups

Carpenter's Math Checklist

Nine simple steps to eliminate most Math mistakes...

1. **Concentrate!**

2. **Draw a Picture...**
   (It helps make sure you're using the right formula.)

3. **Write The Formula...**
   What applies? Rise over Run (Slope), Perimeter, Area, Volume, etc.

4. **Identify the Math Operations you're going to use...**
   (Division, Subtraction, etc.)

5. **Write the Variables...**
   (Usually, these are measurements. *Make sure you convert everything to the same units!*)

6. **Create the Equation...**
   Plug the variables into the formula.

7. **Estimate what you think would be a reasonable answer...**
   Round the variables into more easily manipulated numbers and approximate the result. Check your estimate against your picture. Does it make sense?

8. **Calculate—Write out each step...**
   It doesn't matter if you use a calculator. *Write out each step.*

9. **Check Your Answer—First, check it against your estimate...**
   If you're working a word problem, go back to the question and check the answer. Don't rely on your formula (in case you have picked the wrong one). Check your answer against your picture.

Framing Square Math

Creating a Perpendicular with Thales' Theorem

There are lots of ways to draw right angles from a point on a line. This straightforward method may be the easiest.

You'll Need:
- Pencil
- Compass
- Straightedge

To Start
- We'll start from point B (which is the right hand end of line AB). Mark a point "C" up and to the left of point B (Fig. 1).
- Open your compass to the distance between points B and C.
- Put the point of the compass on C and draw the circle.
- Point A is the intersection on the circle and your original line.
- Use the straight edge to draw a line between points A and C. Extend the line so that it also crosses the circumference of the circle.
- Label this point D.
- Use the straightedge to draw line BD.

BD is perpendicular to AB.
Thank you Thales.

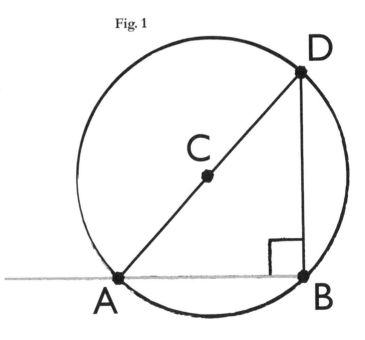

Fig. 1

Framing Square Math

Finding the Height of an Object Using Similar Triangles

You have a very tall tree in your yard and you are thinking about cutting it down. The tree cutters want to know how tall the tree is in order to give you a cost estimate. You do not have a tape measure long enough to measure that tree. How can you determine its height? (Fig. 1)

You'll Need:
- Framing Square
- 25' tape measure (minimum)
- 2' Level
- Plumb bob
- One inch wide wood, plastic or aluminum "sighting stick" approximately three feet long
- Two 1" x 2" wood stakes, 5' to 6' long, each pointed on one end
- Two clamps, either spring clamps or 'C' clamps
- Pencil
- 8 ½" x 11" blank or graph paper.

To Start
- In a line pointing towards the object you're trying to measure (such as the tree), set up the two 1" x 2" stakes approximately 18" to 22" apart by driving the pointed ends approximately 6" into the ground (Figs. 2 & 3).
- Make sure that you have a tape measure long enough to reach from the stakes to the base of your object.
- PLUMB both stakes, both with the level and with the plumb bob. Remember that "Plumb" is 90 degrees to "LEVEL."
- Place the level upright against the stake and watch the bubbles. If the bubbles are not centered, adjust the stake until they are (Fig. 4).
- For the plumb bob, hang it alongside the stake, pointing to the ground. Adjust your stake so it is parallel to the string of the plumb bob (Fig. 5).
- Clamp the square on the plumbed stakes. The tongue (short leg) of the square should point up (Fig. 6).
- Level the body (long leg) of the square.
- Draw a table as shown in Fig. 9.

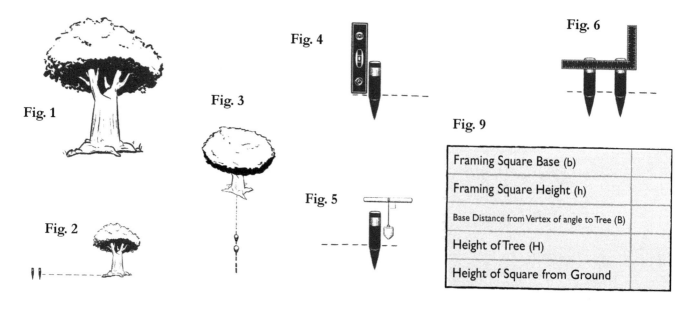

Fig. 1
Fig. 2
Fig. 3
Fig. 4
Fig. 5
Fig. 6
Fig. 9

Framing Square Base (b)	
Framing Square Height (h)	
Base Distance from Vertex of angle to Tree (B)	
Height of Tree (H)	
Height of Square from Ground	

- Place the sighting stick with its lower edge touching the 12" mark on the horizontal leg (body) of the square, and the length crossing the vertical leg (tongue) of the square (Figs. 7 & 8).
- 12" is your Framing Square Base (b). Enter it in your table.
- Bend or kneel so a "sight" can be taken along the bottom edge of the stick to the top of the object being measured. Your eyeball has to be in line with the bottom edge of the stick.
- Note where the bottom edge of the stick crosses the vertical leg of the square. This is Framing Square Height, "h." Enter it in your table.
- Measure the base distance from vertex of your angle to the tree. This is "B." Enter it in your table.

The line of sight creates two similar right triangles, the smaller one on the Framing Square, the larger one with the unknown height and the distance to the tree (Fig. 10).

- Using the data you entered into your table, set up the proportion shown in Figure 11 to find the unknown height (H) of the tree. CROSS MULTIPLICATION generates a simple ALGEBRAIC equation to solve for H.
- Use the Carpenter's Math Checklist (p. 80). Remember to work in the same units. You must convert your measurements into inches, feet, or meters. If you are using a calculator, you must also convert from fractional measurement to decimal measurement.
- Be sure to add the height of the base on which the square is set up to the height of the unknown object!

If two teams of students are taking observations for the same tree and the tree is on a hill, have one team work at the same level as the tree. Have the second work from uphill or downhill of tree. The students results will be different because of the relative heights of their starting points. Have them explain their results and why they are different.

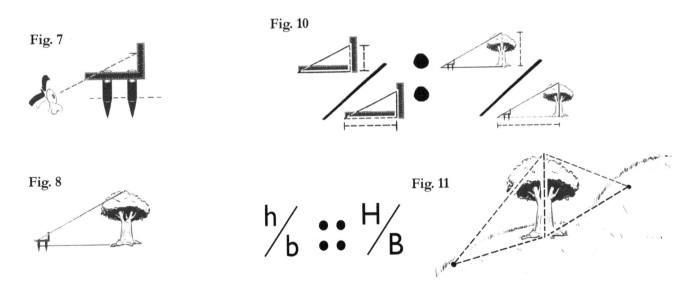

Fig. 7

Fig. 10

Fig. 8

Fig. 11

Framing Square Math

Finding the Height of an Object Using Trigonometry

You have another tall tree in your yard that you are thinking about cutting down (Fig. 1). The tree cutters want to know how tall that tree is in order to give you an estimate. You still do not have a tape measure long enough to measure that tree. This time, how can Trigonometry be used to find the unknown height of an object?

You'll Need:
- Straightedge or 12" ruler
- #2 pencils
- 25' tape measure
- 2 foot level
- Plumb bob
- 1" wide wood, plastic, or aluminum sighting board 3' long
- two 1" x 2" wide wood stakes (5' to 6' long) each pointed at one end
- two clamps (either spring or 'C')
- 24" x 30" sheet of paper
- Framing Square

Remember, Tangent is the ratio in a right triangle found by dividing an angle's **OPPOSITE SIDE** by its **ADJACENT** side (Fig. 2).

To Start:
- Select an object of unknown height, such as the tree (Fig. 1).
- Set up the two 1" x 2" stakes in line, pointed towards the object to be measured (Figs. 3 and 4).
- The stakes should be approximately 18" to 22" apart. Drive the pointed ends of the stakes approximately 6" into the ground. (An easy way to do this is to align the stakes with the shadow of the object.)
- **PLUMB** both stakes with the level (Fig. 5).
- Clamp the Framing Square onto the plumbed stakes. Make sure to **LEVEL** the body (horizontal leg) (Fig. 6).
- The tongue should point up.
- Place the sighting stick with its lower, far corner even with the lower corner of the horizontal leg of the square, and the length crossing the

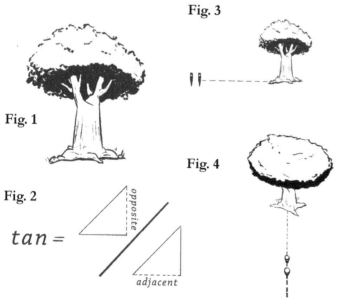

Fig. 3

Fig. 1

Fig. 2

$$tan = \frac{opposite}{adjacent}$$

Fig. 4

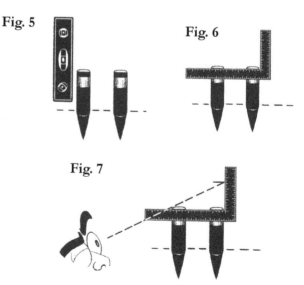

Fig. 5

Fig. 6

Fig. 7

Framing Square Math

vertical leg of the square. Remember to bend or kneel such that a sight can be taken along the bottom edge of the stick to the top of the object being measured (Fig. 7).
- Note the measurements where the bottom edge of the stick crosses the vertical and horizontal legs of the square on a pad (Fig. 8).
- Measure the distance from where the stick crosses the horizontal leg of the square to the base of the tree.
- Remove the square from the stakes.
- Recreate the angle by laying the square on a piece of paper large enough to accommodate it, with the square oriented as it was when the sighting was done.
- Then, draw a line the length of the horizontal leg, and, without moving the square, mark the point at which the sight stick intersected the vertical and horizontal legs.
- Draw a straight line connecting the intersection point with the base point, creating a vertex (Fig. 9).
- Use a **PROTRACTOR** and measure the angle at this vertex.
- Consult a table of Trigonometric functions to find the tangent function for this angle and record the information.

The tangent function equals the **OPPOSITE SIDE** from the angle divided by the **ADJACENT** side.

The distance you measured from where the stick crossed the horizontal leg of the square to the base of the tree is your adjacent side.

- Multiply the adjacent side by the tangent function for your angle and you will derive the length of the opposite side. This is the height of your tree.
- Be sure to add the height of the base on which the square is set up to the height of the unknown object!

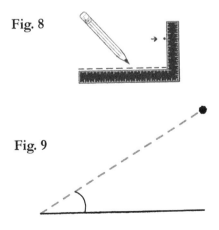

Fig. 8

Fig. 9

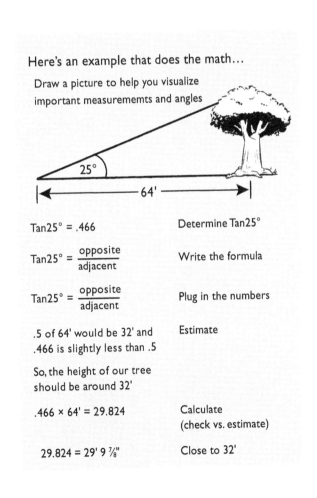

Here's an example that does the math...
Draw a picture to help you visualize important measurememts and angles

Tan25° = .466 — Determine Tan25°

Tan25° = $\frac{opposite}{adjacent}$ — Write the formula

Tan25° = $\frac{opposite}{adjacent}$ — Plug in the numbers

.5 of 64' would be 32' and .466 is slightly less than .5 — Estimate

So, the height of our tree should be around 32'

.466 × 64' = 29.824 — Calculate (check vs. estimate)

29.824 = 29' 9 7/8" — Close to 32'

Framing Square Math

Hands On Math Glossary

Mostly math terms, their meaning, and why they are useful...

Adjacent Side

Definition: Relative to a specific angle in a right triangle, it's the short side (not the hypotenuse) next to the angle.

Application: If you're going to use Trigonometry, you're going to use Adjacent sides. They are a fundamental component of the Trig functions Tangent and Cosine.

Algebra

Definition: A type of math where letters, or symbols, are used as "variables" (instead of specific numbers) to describe relationships that are always true, no matter the size of the individual parts. These relationships can be ratios, proportions, formulas, or equations. The values of the variables can be very small, or very large; but the relationships between them will always be true. Algebra is the "next step up" from Arithmetic when you're calculating with numbers. It's also very closely related to Geometry. They both describe relationships that are true no matter the size of the pieces. Algebra uses numbers. Geometry uses shapes.

Application: Formulas, proportions, data analysis. Algebra is used almost any time a calculator or spreadsheet is used.

Angle Measurement

Definition: Angles get measured in degrees, minutes and seconds—not inches, feet, or meters. Sixty seconds equal one minute. Sixty minutes equal one degree. There are 360 degrees in a circle.

Application: Angles are measured in small and large layout. Small angles happen in the pitch of dovetails. Large angles are measured when laying out the foundations of a building. Tools that measure angles range from precision protractors to laser levels and transits.

Back (Framing Square)

Definition: The side of the square showing when the tongue is to the left and the body is pointing down.

Application: On a high quality framing square, the back usually has the $1/10$th and $1/12$th scales, which greatly increase the usefulness of the tool. The $1/10$ scale is used for decimal calculation; while the $1/12$ scale, which runs along the outside edge of the square's back, turns the square into a 1":1' scale calculator.

Bevel Gauge

Definition: A tool with a body and movable blade that allows you copy an angle.

Application: Small versions are used in cabinet making and boat building for working with angles that aren't 90 degrees. A large version can be built to make computations on a framing square easier.

Body (Framing Square)

Definition: The long leg of the Framing Square.

Glossary

Center

Definition: The point in the middle of the area of a geometric shape, like a rectangle, square, triangle, or circle. The center of a circle is special because it's usually used to draw a circle.

Application: Builders find the center of shapes all the time. Two examples are determining the placement of structural members, or to finding the balance point of an object that is to be lifted. Also, all layout starts from one point. Sometimes that point is the center of a geometric shape.

Circle

Definition: A set of points (which when they're really close together essentially make a line) that are all the same distance (the circle's radius) away from the circle's center. Circles can be classified as a type of conic section—just like squares are classified as a type of rectangle. They are the conic section made by slicing the cone parallel to its base. (A "conic section" is just a slice made through a cone.)

Application: Circles are used throughout the building process. Most drilled holes are circles. Many tables and some windows or buildings are circles. It is a fundamental geometric shape. Parts of circles can be used to make other shapes, like right angles, squares, and isosceles triangles.

Circumference

Definition: The distance around a circle, or any curved shape. It's a circle's perimeter.

Application: Many times the circumference of a circle needs to be determined in order to estimate materials.

Common Denominator

Definition: In fractions, it's a denominator into which two, or more, fractions can be converted. When adding or subtracting fractions, the size of the units have to be the same. This means the denominators of all those fractions need to be the same—or "common."

Application: Used in adding and subtracting measurements, as well as manipulating proportions.

Conic Sections

Definition: Shapes made by slicing a cone. They include: circles, ellipses, parabolas and hyperbolas. Their shapes can also be described by Algebraic formulas. (Algebra and Geometry crossing again.)

Application: Most curves used in construction are conic sections. These curves can be constructed using Geometry, or created by generating coordinates through Algebra.

Coordinate Plane

Definition: A system where a vertical and a horizontal line intersect at a right angle. The vertical line is the Y axis; the horizontal line is the X axis. The point of intersection is called the origin. The intersecting lines form four quadrants. The lines are measured with positive and negative numbers. Above and to the right of the origin are positive measurements. To the left and below the origin are negative numbers.

Application: This system is a graphic way to describe Algebra (lines, slope, conic sections, statistical graphs, etc.). It also is a critical technique for laying out buildings. You have to start from somewhere (the origin.) You then need to measure off of that point to lay out the points that determine the shape of what it is you are building.

Cosecant

Definition: A Trigonometric ratio of a right triangle. In the "Unit Circle," it's the hypotenuse of the large triangle formed with a Cotangent base. It's formula is also the reciprocal of the cosine: 1/cosine. It's not a commonly used function.

Application: When doing large scale layout, it's often necessary to find the length of a triangle's hypotenuse. Knowing all the Trigonometric functions allows a builder to avoid more time consuming "work arounds"—like having to do more layout and then use the Pythagorean Theorem, or again, doing more layout and finding the length through the sine or cosine functions.

Cosine

Definition: A basic Trigonometric ratio of a right triangle's Adjacent side over its Hypotenuse.

Application: Used in carpentry and engineering to layout specific angles and determine distances.

Cotangent

Definition: A Trigonometric ratio of a right triangle. In the "Unit Circle," it's the horizontal, leg of the large triangle formed with a Cosecant hypotenuse.

Application: When doing large scale layout, it's often necessary to find the length of a triangle's leg. Knowing all the Trigonometric functions allows a builder to avoid more time consuming "work arounds"—like having to do more layout and then use the Pythagorean Theorem, or again, doing more layout and finding the length through the sine or cosine functions.

Cross Multiplication

Definition: An incredibly useful technique applicable with two equivalent fractions. If the numerator of the first fraction is multiplied by the denominator of the second and the numerator of the second fraction is multiplied by the denominator of the first, the resulting products will be equal. If you set the problem up as two fractions and an equals sign, the multiplication "crosses" the equals sign.

Application: It's very useful when you know three parts of the problem and need to solve for the fourth. If a problem can be set up as a proportion, it can be solved using this technique.

Cubic Measurement

Definition: Volume measurement. The three dimensions of an object are multiplied by one another to calculate the object's volume. Since there are three dimensions, the exponent used to symbolize cubic measurement is a superscript 3—^{3}.

Application: Estimating volume of anything, such as air or concrete. It is also critical to calculating weight.

Decimal

Definition: A fraction. Really! It's just that the denominator is a "power" of ten. Ten, One Hundred, One Thousand. With the addition of the decimal point, this allows these fractions to be described on the place value chart.

Framing Square Math

Glossary

Application: Decimals are used most times you use a calculator. So, you have to often convert from fractions to decimals and back again; but really, all you're doing is converting from one type of fraction to another.

Decimal Inches

Definition: Fractional inches described in decimals. Since fractions are just division problems, if you solve the problem using long division, you end up with a decimal. ⅛ is 1.000 divided by 8. The solution is .125.

Application: When using an electronic calculator you usually have to convert from fractions to decimal inches, or decimal feet.

Denominator

Definition: In a fraction, the number below the fraction bar that names the number of equal parts (or units) into which the "whole" is divided. If that number is "8," the whole has been divided into eight parts. The word "Denominator" is Latin for "the name"—the name of the unit. The larger the denominator, the smaller the size of the unit. An "eighth" is smaller than a "quarter."

Application: Since the Denominator defines the unit in which you are working, it also defines the tolerance of the work. The more parts in the unit, the finer the tolerance. A carpenter may work in ¹⁄₁₆s, while a machinist may work in ¹⁄₁₀,₀₀₀s. Very different degrees of accuracy—all made possible by the fraction's denominator.

Diagonal

Definition: A line from one corner of a shape to another corner. Diagonals are especially useful in rectangles where they cross at the rectangle's center. And, if the diagonals are the same length, the shape is a rectangle and its corners are square.

Application: This is incredibly useful for checking the accuracy of layout work.

Diameter

Definition: The distance from one edge of a circle to its other edge, passing through the circle's center.

Application: Half of the formula for Pi, diameters are used to calculate the radius, circumference, and areas of a circle, as well as the volume of a sphere.

Dimension

Definition: A distance measured in units such as inches, feet, or meters. There is a difference between a measurement and a dimension when you're talking about applying these terms to existing objects. A measurement is exact. An object has one true measurement. A dimension is by necessity an approximation whose accuracy depends upon the tolerance of your unit of measure: Measuring to a thirty-second of an inch is more accurate than measuring to the nearest half inch.

Application: Dimensions are used all throughout the building process to communicate the size of things, especially when you're building from a set of plans.

Ellipse

Definition: Geometrically, an ellipse is a conic section cut on an angle that doesn't intersect the base. Ellipses can be drawn many ways; but they have one basic property. Once you have the shape laid out, pick a point on its curve. Measure from that point back to the ellipse's two foci. Add those measurements. That total measurement will be the same number no matter which point on the ellipse you pick.

Application: Many things can be built in the shape of an ellipse: table tops, windows, even whole buildings.

Face (Framing Square)

Definition: The side of the square showing when the tongue is to the right and the body is pointing down.

Focal Point/ Foci (plural)

Definition: Foci are the points on the major axis from which you measure to create an ellipse (or any other conic section).

Application: The "pin and string" method of laying out ellipses uses focal points to create the shape. If done with string, or wire, which doesn't stretch, this a good way to draw an ellipse.

Fraction

Definition: A piece of a "whole." It's a relationship set up as a division problem. The division sign that's used is a "fraction bar." There's a number below the bar and a number above the bar. The number below the bar names the number of equal parts (or units) into which the "whole" is divided. If that number is "8," the whole has been divided into eight parts. The official name for this "bottom number" is "Denominator." It's just Latin for "the name"—the name of the unit. The top number of a fraction tells you how many units of the whole are contained in the fraction. The official name for this "top number" is "Numerator." It's just Latin for "numberer"—the number of units. (If we still spoke Latin, fractions would make a lot more sense.)

Application: Fractions are used in measurement. They also form the basis of working with ratios and proportions.

Function

Definition: A relationship written in "math" involving one or more variables where you only get one answer for each set of variables.

Application: On the job, formulas for area and volume are functions. Whenever you use Sine, Cosine, Tangent, or any part of Trig, you're using functions.

Geometry

Definition: A Greek word which means "measuring the Earth." Like Algebra, it describes relationships that are true no matter the size of the pieces; but it uses shapes, rather than numbers.

Application: Geometry is fundamental to the building process. It allows you to determine the location of points in space. This in turn allows a builder to draw the correct lines before they make the correct cuts. An old saying says that: "Carpentry is Geometry."

Hypotenuse

Definition: The longest side of a right triangle—opposite the right angle.

Glossary

Application: Many times the length of a hypotenuse needs to be determined—such as determining the length of roof rafters. Many times a known hypotenuse is used to determine the lengths of the other legs of a triangle, or that triangle's angle.

Improper Fractions

Definition: When the top number (numerator) of a fraction is bigger than the bottom number (denominator). They can be simplified into mixed numbers.

Application: Improper fractions get used many times when multiplying and dividing fractions. This can happen when using formulas or converting measurements.

Inverse

Definition: The word means "Opposite." In fractions, an inverted fraction is one that is flipped. The numerator becomes the denominator, and visa versa. In math operations, an inverse operation is one that reverses the initial operation. Subtraction reverses addition. Multiplication reverses division.

Application: Inverse operations are used to check one another. (Multiplication checks division.) Inverting fractions happens when dividing fractions.

Irregular

Definition: An irregular shape has sides, faces, or angles of differing sizes, while a regular shape has sides, faces, and angles of equal size.

Application: The main use of this word is to communicate clearly on the job. Knowing that a shape is "irregular" means that a builder won't incorrectly try to apply rules that only apply to regular shapes.

Isosceles Triangle

Definition: A triangle where two of its legs are the same length. The base angles of the triangle are the same. And, most importantly for builders, inside the triangle are two symmetrical right angles, which means that it's baseline is perpendicular to its centerline.

Application: Isosceles are critical on the job when creating right angles. If a right angle is created geometrically, there's an isosceles triangle in there somewhere.

Level

Definition: 90 degrees to a "plumb" line. Tangent to the surface of the Earth. A level line is a true "Horizontal."

Application: Level is fundamental to building things on, or in, the ground. Doors swing correctly on their hinges when the top of the door is level and the hinges are plumb.

Linear Equation

Definition: In Algebra, lines and shapes get described by equations. The equation of any straight line is called a linear equation. Linear just means straight. A standard form of the equation is $y = mx + b$. "x" and "y" are the coordinates of a particular point on the line. The variable "m" is the slope of the line and "b" is the y-intercept of the line that is just the point where the line crosses the y axis.

Application: When doing layout, once a line is established, many times it's useful to determine other points that would be on that line. This applies if that line were projected, or if a point within the existing length of the line needed to be determined.

Major Axis

Definition: In an ellipse, it's the long axis. Conventionally, this is usually the horizontal axis.

Application: The line is critical for drawing the correct ellipse. It defines the "length" of the ellipse. The phrase is useful for clearly communicating which line is being discussed.

Measuring/ Measurement

Definition: If it's a verb, it means the act of measuring. If it's a noun, it means the accurate length, width, depth, area, or volume of an object.

Application: Measurement is one of the key skills in building. It's very hard to build well if you can't measure accurately, and manipulate those measurements correctly.

Minor Axis

Definition: In an ellipse it's the shorter, usually vertical, axis.

Application: The line is critical for drawing the correct ellipse. It defines the "height" of the ellipse.

The phrase is useful for clearly communicating which line is being discussed.

Mixed Numbers

Definition: A value made up of whole numbers combined with fractions.

Application: Dimensions from a ruler are usually mixed numbers because they have inches and fractions of an inch.

Opposite Side (see Adjacent Side)

Definition: Relative to a specific angle in a right triangle, it's the short side (not the hypotenuse) opposite the angle.

Application: If you're going to use Trigonometry, you're going to use Opposite sides of right triangles. They are a fundamental component of the Trig functions Tangent and Sine.

Origin

Definition: In the coordinate plane, it's the point where the X and Y axis intersect.

Application: Whenever you lay out a project, you have to start from a point of beginning. Many times, this "Point of Beginning" is an origin where two lines intersect at a right angle.

Parallel

Definition: Describes lines, or planes, that, if they were extended forever, will never touch each other. Another way to say that is that two lines are parallel, if they make the same angle with a third line.

Framing Square Math

Glossary

Application: Parallel lines occur everywhere in the building the process. Sides of rectangles are parallel to one another. Roofs with the same pitch are parallel. Level lines are parallel to one another (as are plumb lines). The floors of a building should be parallel to one another; and they usually should be level.

Perimeter

Definition: The distance around the outside of a shape. In a circle, it's the circumference.

Application: In the building process, figuring perimeter is just as important as figuring the area, or volume, of a shape.

Perpendicular

Definition: Describes lines that form a ninety degree angle. There are many terms that describe the same situation, including: right angle, square and normal.

Application: Even more so than parallel lines, perpendicular lines are everywhere in the building process. Perpendicular lines allow buildings to stand up most efficiently. They create rectangles. And, they are used to create parallel lines.

Pi

Definition: The ratio of a circle's circumference divided by its diameter. The symbol for Pi is π. The formula is $\pi = C/D$. Pi is an irrational number whose approximate values are: 3.14159 (in decimals), or $22/7$ (in fractions).

Application: Pi is used to calculate circumference, diameter, and area.

Plumb

Definition: A line pointing to the center of the Earth is plumb. It is also 90 degrees to level. Such a line can be determined by a spirit level, or by hanging a plumb bob. Some framing squares even have a hole in their tongue on which they will hang with their tongue level and their body plumb.

Application: Another type of line central to the building process. Plumb lines are used to erect walls, hang doors, and set pilings.

Positive and Negative Numbers

Definition: A way to label numbers on a number line and on the coordinate plane. Negative numbers to the left of zero. Positive to the right of zero. On the Coordinate planes, X and Y axis to the right and above the origin.

Application: When doing big layout, such as surveying, positive and negative numbers tell you in which direction you need to move from an existing point in order to establish your next point. This can be for either distances, or for degrees.

Proportion

Definition: Two equivalent ratios create a proportion. They are frequently written as fractions. The symbol for a proportion is ::

Application: When a given ratio is known, proportions can usually be used to calculate any needed, related, information. Certainly anything with "rise and run," such as stairs or roofs, involve proportions. They are the numerical equivalent of similar right triangles.

Protractor

Definition: A tool used to measure angles in degrees.

Application: Most protractors are small. So, they're used to measure small angles in furniture or precision work.

Pythagorean Theorem

Definition: A formula that describes the relationship between the lengths of the sides of any right triangle:
$a^2 + b^2 = c^2$

Application: Finding the side lengths of right triangles allows builders to use math to "predict the future." For example, needed materials can be calculated and gathered before pieces get built. This applies to carpenters calculating the length of rafters, or a boat rigger determining the length of the stays necessary to hold up a mast.

Ratio

Definition: A relationship between two numbers. Two equivalent ratios make a proportion.

Application: Ratios happen everywhere in building. Rise over Run is probably the most common building ratio.

Rectangles

Definition: Four-sided shapes where: the opposite sides are the same length straight lines. The opposite sides are parallel to one another. All four corners are right angles. And, the diagonals are the same. A square is a rectangle with all four sides the same length.

Application: Most buildings and much furniture is laid out by creating rectangles. They are the most common geometric shape in the building process.

Right Angles

Definition: Angles that contain 90 degrees, Other terms that describe the same situation include: square, perpendicular, and normal.

Application: It's so important to building that it's the "right" angle. Every other angle is wrong—at least they're not so useful in building. Right angle's align loads and help buildings stand up. They allow us to hang doors that work well. And, they look "right" to our eye. So, they are a critical component of design, as well as engineering and building.

Rise and Run

Definition: A vertical distance divided by a horizontal distance, Rise and Run is the most common ratio in building. It describes "Slope" in Algebra. It also describes the Tangent function in Trigonometry.

Application: Wherever proportional triangles occur, there's "rise over run." Its uses range from carpentry (stair building and roof framing), to millwrighting (determining coupling gap in huge powertrains.) The more you look, the more examples you'll find.

Rise

Definition: Vertical distance in a Rise and Run ratio.

Application: One example of thousands: The height of a stair riser.

Framing Square Math

Glossary

Run

Definition: The horizontal distance in a Rise and Run Ratio

Application: One example of thousands: The depth of a stair tread.

Scales (Framing Square)

Definition: The rulers along the front and back edges of the square.

Secant

Definition: A Trigonometric ratio that is the reciprocal of the sine function. It's the hypotenuse of the "Sine, Cosine" Trig triangle.

Similar Triangles

Definition: Triangles that have all the same angles. The lengths of the triangles' legs are proportional.

Application: Similar triangles are the mathematical "power" behind using the square as a calculator.

Sine

Definition: A Trigonometric ratio of opposite over hypotenuse.

Slipping the Square

Definition: Manipulating the square while maintaining its angle to an external straightedge, or slipping line.

Application: This process develops proportions that are used to make calculations.

Slope

Definition: Rise over Run. Answers "How steep is the line?"

Application: Slope is involved in almost every building trade: pipe drainage, roof pitch, and glue joints all depend on slope.

Square Measurement

Definition: Area measurement. Area is measured in two dimensions and is therefore "square."

Application: Square measurement is used to estimate materials in almost every building process, such as drywall for walls, plywood for a floor, or paint for a room.

Square

Definition: A rectangle with 4 equal sides. Also, a tool that creates a right angle. Also, 90 degrees. Also, "squared" units are used to measure area.

Symmetry

Definition: Two identical shapes on either side of a line, or "axis of symmetry." A mirror image.

Application: Many times a structure or an object needs to be identical on either side of a center line. This is especially true for boats.

Tangent

Definition: A Trigonometric ratio of opposite over adjacent.

Application: The most applied Trigonometric function in the building process because it is "Rise over Run."

Thales' Theorem

Definition: When the corner of a right angle (its vertex) is placed on a circle's circumference and its legs are pointing into the circle, the legs of the right angle will cross the circumference of the circle at one of the circle's diameters.

Application: This can be used to find the center of a circle, draw circles, or to create right angles.

Tolerance

Definition: The degree of accuracy to which a problem is worked, or an object is built.

Application: Deciding on the correct level of tolerance for the job and for the materials being used to build, is one of the signs of a craftsperson. Anybody can spend too much time over-calculating a math problem, or over-thinking a building problem.

Tongue (Framing Square)

Definition: The short leg of the square.

Trammel

Definition: A tool "that restricts, or guides, movement."

Application: In our case, it's a gauge that guides the movement of a pencil to draw an ellipse. Trammel points are tools that attach to a piece of wood to make a beam compass. They restrict the movement of the compass to draw specific circles.

Trigonometric Functions

Definition: Ratios of the sides of triangles with unique values for specific angles.

Application: They can be used to accurately determine angles and distances in almost any building situation.

Trigonometry

Definition: Literally means "measurement of triangles." In our case, it's the measurement of right triangles that results in the Trigonometric functions.

Application: Enables someone to determine lengths and angles to high precision. Generally used in large layout, such as surveying, or precision layout, such as machinery alignment.

Vertex

Definition: The corner of an angle.

Application: Accurate vocabulary for clear communication.

Whole Numbers

Definition: "Integers" in Latin.

Y Intercept

Definition: Where a line crosses the y axis in the coordinate plane. Formula component that locates a line on a layout.

CPSIA information can be obtained
at www.ICGtesting.com
Printed in the USA
FSHW020748270420
69535FS